OXFORD MATHEMATICAL MONOGRAPHS

Editors

G. HIGMAN   G. TEMPLE

PARTIALLY ORDERED TOPOLOGICAL VECTOR SPACES

BY

YAU-CHUEN WONG

AND

KUNG-FU NG

CLARENDON PRESS · OXFORD
1973

Oxford University Press, Ely House, London W. 1

GLASGOW NEW YORK TORONTO MELBOURNE WELLINGTON
CAPE TOWN IBADAN NAIROBI DAR ES SALAAM LUSAKA ADDIS ABABA
DELHI BOMBAY CALCUTTA MADRAS KARACHI LAHORE DACCA
KUALA LUMPUR SINGAPORE HONG KONG TOKYO

ISBN 0 19 853523 6

PRINTED IN NORTHERN IRELAND
AT
THE UNIVERSITIES PRESS, BELFAST.

TO OUR TEACHERS

J. D. Weston

S. T. Tsou

H. P. Rogosinski

PREFACE

The duality theory is one of the most important and fruitful theories in the study of topological vector spaces. This book gives an account of the duality theory of *partially ordered* topological vector spaces. Through the pioneering works of many great mathematicians including Riesz, Frendenthal, Birkhoff, Kakutani, Kantorovitch, Krein, and Nakano, the theory of Riesz spaces (vector lattices), and in particular the theory of Banach lattices, has been well developed; there are many excellent books containing information on the classical theory of Banach lattices (for example, Day (1962), Kelley-Namioka (1963)). The theory was further developed by Luxemburg and Zaanen through their systematic and extensive studies carried out in a series of papers entitled 'Notes on Banach Function Spaces' published in the 60's.

The classical duality theory of Banach lattices has been generalized and developed in the following two divergent directions:

(a) Partially ordered Banach spaces. Krein seems to be the man who initiated this subject; in particular, he shows how the order properties of the Banach space E determine the dual order properties of the Banach dual space E'. Andô, Edwards, and Ellis initiated the attack on the (much more difficult) converse problem; they show that some of the order properties of E are completely determined by the dual properties of E'. Studies of this kind have been carried further by Davies, Ng, Asimow, Perdrizet, and others. The development of the abstract theory of partially ordered Banach spaces has much been influenced and motivated by other branches of mathematics (e.g. C^*-algebra theory and Choquet boundary theory); on the other hand, the abstract theory has also enriched our understanding in other branches (cf. for example, the works of Effros, Størmer, Alfsen, and others).

(b) Locally convex Riesz spaces. The study of such spaces has been strongly influenced by the general theory of topological vector spaces and abstract integration theory (Köthe spaces theory in particular). Roberts seemed to be the first to investigate the duality theory for locally convex Riesz spaces. The theory has then been developed rapidly through the works of Namioka, Schaefer, Kawai, Peressini, Goffman, Wong, and Fremlin. In most of their investigations, a remarkable theorem of Nakano (asserting that for topological

Riesz spaces topological completeness follows from certain order completeness assumptions) plays an important role. By utilizing this powerful theorem, Schaefer proved in 1960 that each reflexive locally convex Riesz space is topologically complete. This result is remarkable in view of Kōmura's example of a non-complete locally convex reflexive space, and indicates that order properties may substantially 'influence' the topological structure. In 1969, Wong extended Schaefer's theorem to semi-reflexive spaces by considering the Dieudonné topology σ_S (and applying Nakano's theorem); he also showed that σ_S is relevant for establishing a converse of Nakano's theorem.

The present book is an attempt to provide a unifying and balanced treatment of the above two seemingly unrelated subjects (a) and (b). Accordingly, the book is roughly divided into three parts. The first part consisting of the first eight chapters is mainly on the theory of partially ordered locally convex spaces, that is, the 'common' theory which is applicable to both (a) and (b). Apart from the work of those mentioned above, we have also included some work by Bonsall and by Weston, and some recent results of Jameson and Duhoux. The second part consists of one single (long) chapter and is a brief account of the results described in (a) with emphasis on those duality results peculiar to normed spaces. The third part consists of the last nine chapters, in which is given not only the theory described in (b) but also Wong's work on barrelled, order-infrabarrelled, infrabarrelled, and bornological locally convex Riesz spaces.

At the end of the book we have included 'Notes on the Bibliography'. These, together with some comments and remarks distributed through the text, should be sufficient to lead the reader to the relevant literature. In most cases we have not attempted to trace the origins of the results but instead to refer the reader to standard reference books whenever possible. Thus, for example, when we say that certain results are due to Schaefer (cf. Schaefer (1966)), we mean that the results and some related material can be found in his book published in 1966 even though he published his results in a much earlier paper. Likewise, when we say that by Schaefer (1966, p. 126), we mean 'by a result appearing on page 126 of his book,' even though the result may not be due to Schaefer.

The book has evolved from lecture notes prepared by the authors for seminars conducted in the Chinese University of Hong Kong. These seminars were attended primarily by advanced undergraduated students (in their fourth year) and also by some of our colleagues. We believe that this book should be quite accessible to any student who is acquainted with the elementary theory of topological vector spaces.

The authors' interest in the subject was cultivated when they did their research at University College, Swansea. It gives them great pleasure to thank Professor J. D. Weston, the Department Head, and his staff members, in particular Dr A. J. Ellis, Dr H. P. Rogosinski, and Professor G. M. Peterson (who is now at the University of Canterbury) for their guidance and stimulation. We would also like to thank Professor Weston for suggesting that we write the book and for his help in arranging for its publication. The manuscript was written in Hong Kong; we should like to take the opportunity to thank the Chinese University of Hong Kong in general and the United College in particular for financial and moral support. We wish in particular to thank Dr S. T. Tsou who constantly encourages and helps us in many ways, not only as the Department Head but also as a teacher and as a friend. We are also grateful to the staff of the Clarendon Press for their assistance.

United College   Y. C. W.
The Chinese University of Hong Kong   K. F. N.
May 1973

CONTENTS

1

FUNDAMENTALS OF
ORDERED VECTOR SPACES

In this chapter we review some basic facts in linear algebra and ordered vector spaces, which we shall need in what follows. The following notation is used: ϕ denotes the *empty set*, and $A\backslash B$ denotes the *complement* of B relative to A, where A and B are sets.

Throughout this book we shall restrict our attention to vector spaces over the real field **R**. If A and B are subsets of a vector space E and if λ, μ are real numbers, we define

$$\lambda A + \mu B = \{\lambda a + \mu b : a \in A, \, b \in B\}.$$

The expression $\{x\} + A$ will be abbreviated by $x + A$, $(-1)A$ by $-A$, and $A + (-B)$ by $A - B$. Let K be a subset of E. K is said to be *convex* if $\lambda K + (1-\lambda)K \subseteq K$ whenever $0 \leqslant \lambda \leqslant 1$, K is said to be *symmetric* if $-K = K$, and K is *circled* if $\lambda K \subseteq K$ whenever $|\lambda| \leqslant 1$. If B is a subset of E, the smallest convex set containing B, denoted by co B, is called the *convex hull* of B; and the smallest convex circled set containing B (denoted by ΓB) is called the *convex circled hull* (or *absolutely convex hull*) of B. Let A and B be subsets of E. We say that A *absorbs* B if there exists $\lambda > 0$ such that $B \subseteq \mu A$ for all μ with $|\mu| \geqslant \lambda$. If A absorbs every finite subset of E, then A is said to be *absorbing*. If A is absorbing, the functional p_A defined by

$$p_A(x) = \inf\{\lambda > 0 : x \in \lambda A\} \quad \text{for any } x \text{ in } E$$

is called the *gauge* (or *Minkowski functional*) of A. A functional p on E is said to be *sublinear* if

$$p(x+y) \leqslant p(x) + p(y) \quad \text{and} \quad p(\lambda x) = \lambda p(x)$$

for all x, y in E and $\lambda \geqslant 0$. A sublinear functional p on E is called a *semi-norm* if $p(\mu x) = |\mu|\, p(x)$ for all x in E and μ in **R**.

A non-empty convex subset C of E is called a *cone* if $\lambda C \subseteq C$ for all $\lambda \geqslant 0$. Clearly a cone C in E determines a *transitive* and *reflexive relation* '$\leqslant$' by

$$x \leqslant y \quad \text{if} \quad y - x \in C;$$

moreover this relation is compatible with the vector structure, i.e.

(a) if $x \geqslant 0$ and $y \geqslant 0$ then $x+y \geqslant 0$,

(b) if $x \geqslant 0$ then $\lambda x \geqslant 0$ for all $\lambda \geqslant 0$.

The relation determined by the cone C is called the *vector (partial) ordering* of E, and the pair (E, C) (or $(E, \leqslant)$) is referred to as an (*partially*) *ordered vector space*. Conversely if ' $\leqslant$ ' is a relation in E which is transitive, reflexive, and compatible with the vector structure of E and if we define

$$C = \{x \in E : x \geqslant 0\},$$

then C is a cone in E, and ' $\leqslant$ ' is exactly the vector ordering of E induced by C.

A cone C in E is said to be *proper* if $C \cap (-C) = \{0\}$. The vector ordering ' $\leqslant$ ' of E, induced by a cone C, is *antisymmetric* if and only if C is proper.

It is easily seen that the intersection of a family of cones in E is a cone. The smallest cone containing a given set A is denoted by pos A. Clearly

$$\text{pos } A = \left\{ \sum_{i=1}^{n} \lambda_i a_i : a_i \in A, \ \lambda_i > 0 \quad \text{for all } i = 1, 2, ..., n \right\}.$$

We see that pos A is proper if and only if all $\lambda_i = 0$ $(i = 1, 2, ..., n)$ whenever $\sum_{i=1}^{n} \lambda_i a_i = 0$, where $a_i \in A \backslash \{0\}$ and $\lambda_i > 0$ $(i = 1, 2, ..., n)$. If A is convex, then pos A has a simpler expression as shown in the following proposition.

(1.1) PROPOSITION. *If A is a non-empty convex subset of a vector space E, then*

$$\text{pos } A = \cup \{\lambda A : \lambda \geqslant 0\}.$$

Furthermore, if $0 \notin A$ then pos A is proper.

Proof. Suppose that

$$P = \cup \{\lambda A : \lambda \geqslant 0\}.$$

It is clear that $\mu P \subseteq P$ for all $\mu \geqslant 0$. From the convexity of A, it is easy to show that P is convex, and so P is a cone containing A. Further, if W is a cone containing A then

$$\lambda A \subseteq W \quad \text{for all } \lambda \geqslant 0,$$

and so $P \subseteq W$. Therefore P is the smallest cone containing A, i.e. pos $A = P$.

Finally, we show that pos A is proper whenever $0 \notin A$. Suppose, on the contrary, that there exist a, b in A and λ, $\mu > 0$ such that

$$\lambda a = -\mu b.$$

Then the convexity of A entails that

$$0 = \frac{\lambda}{\lambda + \mu}\, a + \frac{\mu}{\lambda + \mu}\, b \in A,$$

which gives a contradiction. This completes the proof.

A cone C in E is said to be *generating* if $E = C - C$.

(1.2) PROPOSITION. *Let C be a cone in E. Then the following statements are equivalent:*

(a) *C is generating;*

(b) *for any $x \in E$, there exists $u \in C$ such that $u \geqslant x$;*

(c) *the vector ordering in E is directed in the sense that for any x, y in E there exists $z \in E$ such that $x \leqslant z$ and $y \leqslant z$.*

Proof. Straightforward.

Let $(E, \leqslant)$ be an ordered vector space. The vector ordering $\leqslant$ is said to be *Archimedean* if $x \leqslant 0$ whenever $nx \leqslant y$, or *almost-Archimedean* if $x = 0$ whenever $-y \leqslant nx \leqslant y$, for all positive integers n and some $y \in E$. It should be noted that if the vector ordering is proper and Archimedean then it is almost-Archimedean, but the word 'proper' cannot be dropped in the above conclusion as shown by the following examples. (a) In $\mathbf{R}^2$, suppose that

$$C = \{(x, y) : x \geqslant 0\}$$

and that '$\leqslant$' is the vector ordering determined by the (improper) cone C. Then '$\leqslant$' is Archimedean, but not almost-Archimedean.

(b) $\{(\alpha, \beta) \in \mathbf{R}^2 : \alpha > 0,\ \beta > 0\} \cup \{(0, 0)\}$ is a proper almost-Archimedean cone in $\mathbf{R}^2$, but it is not Archimedean.

In the definition of almost-Archimedean ordering, the element y involved in the inequalities $-y \leqslant nx \leqslant y$ must be positive (that is, in the positive cone of E). In the definition of Archimedean ordering we can also consider only positive elements y. More precisely, if $(E, \leqslant)$ satisfies the property that $x \leqslant 0$ whenever $nx \leqslant y'$ for all positive integers n and some $y' \in E$ with $y' \geqslant 0$, then '$\leqslant$' is Archimedean. (In fact, suppose $nx \leqslant y$ for all positive integers n and for some $y \in E$, where y is *not* necessarily positive. Let $y' = y - x$. Then $y' \geqslant 0$ and $nx \leqslant y'$.) This observation makes the following two propositions clear.

(1.3) PROPOSITION. *For any ordered vector space* $(E, \leqslant)$, *the following statements are equivalent:*

(a) *the vector ordering '$\leqslant$' is Archimedean;*

(b) *if x, y in E are such that $x \leqslant \lambda y$ for all $\lambda > 0$ then $x \leqslant 0$;*

(c) *if x, y are in E and ε is a positive real number such that $x \leqslant \lambda y$ for all λ with $0 < \lambda < \varepsilon$, then $x \leqslant 0$.*

(1.4) PROPOSITION. *For any ordered vector space* $(E, \leqslant)$, *the following statements are equivalent:*

(a) *the vector ordering $\leqslant$ is almost-Archimedean;*

(b) *if x, y are in E such that $-\lambda y \leqslant x \leqslant \lambda y$ for all $\lambda > 0$ then $x = 0$;*

(c) *if x, y, z are in E and ε is a positive real number such that $\lambda y \leqslant x \leqslant \lambda z$ for all λ with $0 < \lambda < \varepsilon$, then $x = 0$.*

Let F be a vector subspace of an ordered vector space (E, C), where C is a cone in E. Then $F \cap C$ is a cone in F and the vector ordering induced by $F \cap C$ is called the *relative ordering*. The ordering for a subspace will always be assumed to be defined in this manner. It should be noted that if C is a proper cone then so is $F \cap C$. However, it may happen that $F \cap C$ is a proper cone while C is not. It is clear that the relative ordering of an Archimedean ordering is again Archimedean and the relative ordering of an almost-Archimedean ordering is almost-Archimedean.

Examples

(a) The usual ordering for **R** is that induced by the proper cone **R**$^+$ of all non-negative real numbers.

(b) If E is a vector space of real-valued functions defined on a set S then the usual ordering for E is defined pointwise:

$$f \leqslant g \text{ in } E \Leftrightarrow f(s) \leqslant g(s) \quad \text{for all } s \text{ in } S.$$

In particular, if S is a topological space then the ordering for the space of all continuous real-valued functions on S is defined in this manner. Another example of pointwise-defined ordering is that in sequence spaces.

(c) The usual ordering for the vector space ω of all sequences $\{\lambda_n\}$ of real numbers is that induced by the cone K consisting of all sequences

$\{\lambda_n\}$, where each $\lambda_n \geqslant 0$. The following subspaces of ω are endowed with the relative ordering:

m: the space of all bounded sequences of real numbers;

c: the space of all convergent sequences of real numbers;

c_0: the space of all null sequences (that is, sequences converging to zero);

l^p: the space of all l^p-summable sequences of real numbers.

(d) Let (Ω, μ) be a measure space and E the vector space of all measurable real-valued functions. Here, as usual, functions which are equal almost everywhere are identified. For f, g in E, we define

$$f \leqslant g \Leftrightarrow f(x) \leqslant g(x) \quad \text{for almost every } x \text{ in } \Omega.$$

The subspace of all L^p-summable functions in E is endowed with the relative ordering, where p is a positive real number.

In all the examples (a)–(d), the vector orderings are Archimedean.

Let B be a subset of an ordered vector space (E, C). B is said to be *majorized* if there exists x in E such that $b \leqslant x$ for all b in B; B is said to be *minorized* if there exists x in E such that $x \leqslant b$ for all b in B; B is said to be *positive* if $B \subseteq C$. For a pair of elements x, y in E with $x \leqslant y$, let

$$[x, y] = \{z \in E : x \leqslant z \leqslant y\}.$$

The sets of the form $[x, y]$ are called *order-intervals*. It is easily verified that order-intervals are convex; the converse is, of course, inexact. A subset in (E, C) is said to be *order-bounded* if it is contained in some order-interval in E. A subset A of E is said to be *order-convex* (or *full*) if $[a_1, a_2] \subseteq A$ whenever a_1, $a_2 \in A$ and $a_1 \leqslant a_2$. An order-convex set is not necessarily convex. For instance, consider $\mathbf{R}^2$ with the cone C defined by

$$C = \{(\alpha, \beta) \in \mathbf{R}^2 : \alpha \geqslant 0, \beta \geqslant 0\}.$$

Then $\mathbf{R}^2 \backslash C$ is order-convex but not convex. In fact, we have the following proposition.

(1.5) PROPOSITION. *Let (E, C) be an ordered vector space. Then the following statements are equivalent:*

(a) *each order-convex set in E is convex;*

(b) *the vector ordering induced by C is total, that is, $E = C \cup -C$.*

Proof. (a) $\Rightarrow$ (b): Clearly $E \backslash C$ is order-convex, so must be convex by (a). If C is not total in E, let $x \in E \backslash (C \cup -C)$. Then both x and $-x$

2

are in the convex set $E \backslash C$; hence

$$0 = \tfrac{1}{2}x + \tfrac{1}{2}(-x) \in E \backslash C,$$

contrary to the fact that the cone C contains 0.

(b) $\Rightarrow$ (a): Let B be an order-convex set. Let $b_1, b_2 \in B$, and $\lambda_1, \lambda_2 > 0$ with $\lambda_1 + \lambda_2 = 1$. Since the vector ordering in E is total, we have either $b_1 \leqslant b_2$ or $b_1 \geqslant b_2$. Without loss of generality, suppose $b_1 \leqslant b_2$. Then $b_1 \leqslant \lambda_1 b_1 + \lambda_2 b_2 \leqslant b_2$, so $\lambda_1 b_1 + \lambda_2 b_2 \in [b_1, b_2] \subseteq B$ by the order convexity of B.

Simple examples in the plane show also that convex sets are not necessarily order-convex. Thus the concepts of convexity and order convexity are quite distinct. A subset A in an ordered vector space is said to be *o-convex* if it is both convex and order-convex. The o-convexity must not be confused with the order-convexity.

It is clear that the intersection of order-convex sets is order-convex and the intersection of o-convex sets is o-convex. Given a subset B in (E, C), the smallest order-convex set in E containing B is called the *order-convex hull of B* and is denoted by $[B]$. Notice that

$$[B] = \{x \in E : b_1 \leqslant x \leqslant b_2 \text{ for some } b_1, b_2 \text{ in } B\} = (B+C) \cap (B-C).$$

The order-convex hull $[B]$ will sometimes be referred to as the *full hull* of B and will be alternatively denoted by $F(B)$.

A subset A of (E, C) is called:

(a) *positive-order-convex* if $[0, a] \subseteq A$ whenever $a \in A$ and $a > 0$ (thus A contains 0 if it contains a positive element);

(b) *absolute-order-convex* if $[-a, a] \subseteq A$ whenever $a \in A$ and $a \geqslant 0$. If A is a symmetric set then the order-convexity implies (b) which in turn implies (a). Simple examples in the plane show that the three order-convexities are distinct.

A semi-norm p on (E, C) is called:

(a′) *monotone* if $p(x) \leqslant p(y)$ whenever $0 \leqslant x \leqslant y$ in E;

(b′) *absolute-monotone* if $p(x) \leqslant p(y)$ whenever $-y \leqslant x \leqslant y$ in E.

The following result demonstrates the relationship between these concepts.

(1.6) PROPOSITION. *Let p be a semi-norm on an ordered vector space (E, C) and let $V = \{x \in E : p(x) < 1\}$. Then the following statements hold:*

(a) *p is absolute-monotone if and only if V is absolute-order-convex;*

(b) *p is monotone if and only if V is positive-order-convex;*

(c) *V is order-convex if and only if p satisfies the following implication:*

$$x \leqslant y \leqslant z \text{ in } E \Rightarrow p(y) \leqslant \max\{p(x), p(z)\}. \qquad (1.1)$$

(If V is replaced by the closed ball $\Sigma = \{x \in E : p(x) \leqslant 1\}$, the proposition remains true.)

Proof. Straightforward.

(If condition (1.1) is satisfied, C will be referred to as a *1-normal cone* in E with respect to p.)

An important concept dual to the order-convexity is that of decomposable sets. A set A in an ordered vector space (E, C) is said to be *decomposable* if for each a in A there exist a_1, a_2 in $A \cap C$ such that $a = \lambda_1 a_1 - \lambda_2 a_2$ for some $\lambda_1, \lambda_2 \geqslant 0$ with $\lambda_1 + \lambda_2 = 1$. Note that a symmetric, decomposable set may not be convex and that a symmetric convex set is not necessarily decomposable. Observe, however, that the symmetric convex hull of a decomposable set is decomposable. Also the union of a family of decomposable sets is decomposable, but the intersection of decomposable sets may not be decomposable.

Given any set A in (E, C), the union of all decomposable sets in E contained in A, which is the largest decomposable set contained in A, will be called the *decomposable kernel* of A, and will be denoted by $D(A)$. If A is circled and convex then $D(A)$ can be explicitly expressed as

$$D(A) = \{\lambda_1 a_1 - \lambda_2 a_2 : a_1, a_2 \in A \cap C, \lambda_1, \lambda_2 \geqslant 0, \lambda_1 + \lambda_2 = 1\}$$
$$= \Gamma(A \cap C) = \mathrm{co}(-(A \cap C) \cup (A \cap C)).$$

A subset B in (E, C) is said to be *positively generated* if for each b in B there exist b_1, b_2 in $B \cap C$ such that $b_1 - b_2 = b$. Thus E itself is positively generated if C is a generating cone. It is also clear that any symmetric convex and decomposable set is positively generated.

(1.7) PROPOSITION. *Let p be a semi-norm on an ordered vector space (E, C) and let $V = \{x \in E : p(x) < 1\}$. Then the following statements are equivalent:*

(a) *V is decomposable;*

(b) *for each x in E and each positive real number ε there exist two positive elements x_1, x_2 with $p(x_1) + p(x_2) \leqslant p(x) + \varepsilon$ such that $x = x_1 - x_2$.*

Proof. (a) $\Rightarrow$ (b): Note that $x \in (p(x) + \varepsilon)V$. Since V is decomposable, there exist v_1, v_2 in $V \cap C$ and $\lambda_1, \lambda_2 \geqslant 0$ with $\lambda_1 + \lambda_2 = 1$ such that

$$x = (p(x) + \varepsilon)(\lambda_1 v_1 - \lambda_2 v_2).$$

Let $x_1 = \lambda_1(p(x) + \varepsilon)v_1$ and $x_2 = \lambda_2(p(x) + \varepsilon)v_2$. Then (b) is easily verified.

(b) $\Rightarrow$ (a): Let $v \in V$. Then there exists $\varepsilon > 0$ such that $p(v) + \varepsilon < 1$. For these v and ε, by (b), there exist $x_1, x_2 \in C$ with

$$p(x_1) + p(x_2) < p(v) + \varepsilon < 1$$

such that $v = x_1 - x_2$. Take positive real numbers δ_1, δ_2 such that $0 < p(x_i) < \delta_i$ for each i, and $\delta_1 + \delta_2 = 1$. Then x_1/δ_1 and x_2/δ_2 are in V and $v = \delta_1(x_1/\delta_1) - \delta_2(x_2/\delta_2) \in D(V)$. Therefore V is decomposable.

A subset B of (E, C) is said to be:

(i) *absolutely dominated* if for each b in B there exists x in B such that $-b, b \leqslant x$;

(ii) *positively dominated* if for each b in B there exists x in B such that $0, b \leqslant x$.

Clearly the condition (i) implies (ii). Also a decomposable convex set A is always absolutely dominated. In fact, if $a = \lambda_1 a_1 - \lambda_2 a_2$ where $a_1, a_2 \in A \cap C$, $a \in A$, $\lambda_1, \lambda_2 \geqslant 0$, and $\lambda_1 + \lambda_2 = 1$, then let

$$x = \lambda_1 a_1 + \lambda_2 a_2;$$

since A is convex, $x \in A$. It is clear that $\pm a \leqslant x$. Simple examples show that an absolutely dominated and convex set may not be decomposable. A set which is both absolutely dominated and absolute order-convex is said to be *solid*. Solid sets may not be convex and are not necessarily decomposable.

Example. Let $\mathbf{R}^2$ be the plane with the usual ordering ($(\alpha, \beta) \geqslant 0$ if and only if $\alpha \geqslant 0$ and $\beta \geqslant 0$). Let $A = \{(\alpha, \beta) \in \mathbf{R}^2 : |\alpha| \leqslant 2, |\beta| \leqslant 2\}$ and $B = \{(\alpha, \beta) : |\alpha| + |\beta| \leqslant 3\}$. Then A is a non-decomposable solid set, and B is a solid set but not order-convex. The union $A \cup B$ is solid but neither circled nor convex nor order-convex.

The proof of the following proposition is similar to that given for the preceding proposition and therefore will be omitted.

(1.8) PROPOSITION. *Let p be a semi-norm on an ordered vector space (E, C) and let $V = \{x \in E : p(x) < 1\}$. We consider the following statements:*

(a) *V is absolutely dominated;*

(a') *for each x in E and each $\varepsilon > 0$ there exists y in E with*

$$p(y) < p(x) + \varepsilon$$

such that $-x, x \leqslant y$;

(b) *V is positively dominated;*

(b') *for each x in E and each $\varepsilon > 0$ there exists y in E with*

$$p(y) \leqslant p(x) + \varepsilon$$

such that $0, x \leqslant y$.

Then the statements (a) *and* (a') *are equivalent, and statements* (b) *and* (b') *are equivalent.*

An ordered vector space (E, C) is said to have the *Riesz decomposition property* if $[0, u] + [0, v] = [0, u + v]$ whenever u, v are elements in C. It is easily verified that (E, C) has the Riesz decomposition property if and only if $[x_1, y_1] + [x_2, y_2] = [x_1 + x_2, y_1 + y_2]$ for all x_i, y_i in E with $x_i \leqslant y_i$ for each i. The following proposition is well known; the reader is referred to Bourbaki (1965) or Fuchs (1966) for its proof.

(1.9) PROPOSITION. *For an ordered vector space (E, C) the following statements are equivalent*:

(a) (E, C) *has the Riesz decomposition property;*

(b) *if* x_i, $y_j \in E$ *and* $x_i \leqslant y_j$ *for* $i = 1, 2, \ldots, n$, *and* $j = 1, 2, \ldots, m$, *there exists z in E such that* $x_i \leqslant z \leqslant y_j$ *for all i, j;*

(c) *if* x_i, $y_j \in C$ *for all* $i = 1, 2, \ldots n$ *and* $j = 1, 2, \ldots m$, *and if* $\sum_{i=1}^{n} x_i = \sum_{j=1}^{m} y_j$ *then there exist $c_{ij} \in C$ such that* $x_i = \sum_{j=1}^{m} c_{ij}$ *and* $y_j = \sum_{i=1}^{n} c_{ij}$.

A *vector lattice* (*Riesz space*) is defined to be an ordered vector space (E, C) with *proper* cone C such that, for any pair of elements x, y in E, sup (x, y) (the supremum of x and y) and inf(x, y) (the infimum of x and y) exist in E. It is well known and elementary that each vector lattice has the Riesz decomposition property, but the converse is false.

Next let us turn to a discussion of 'dual orderings'. A functional f (not necessarily linear) on an ordered vector space (E, C) is said to be *order-bounded* if f is bounded on each order-bounded subset of E and is said to be *positive* if $f(x) \geqslant 0$ for all $x \in C$. Let E^* denote the algebraic dual of E consisting of all linear functionals on E. Let E^b denote the set of all order-bounded linear functionals on E and C^* the set of positive linear functionals on E. Then $C^* \subseteq E^b \subseteq E^*$, and C^* is a cone in E^* while E^b is a vector subspace of E^*. The space E^b will be conveniently referred to as the *order-bound dual* of E. C^* will be called the *dual cone* in E^* and the induced vector ordering will be referred to as a *dual ordering*. Unless an explicit statement is made to contrary, the ordering in E^* will always be the one induced by C^*, and the ordering in any subspace F of E^* will be the relative ordering induced by $C^* \cap F$. In particular, the order-bound dual E^b has the dual ordering induced by $C^* \cap E^b = C^*$. Let $E^\#$ denote the linear hull of C^* in E^*, that is, $E^\# = C^* - C^*$. Then $E^\# \subseteq E^b \subseteq E^*$, where the inclusions may be strict (cf. Namioka (1957)). The ordered vector space $(E^\#, C^*)$ will be referred to as the *order dual* of E.

(1.10) THEOREM (Riesz).   *Let (E, C) be an ordered vector space with the Riesz decomposition property and $E = C - C$. Then (E^b, C^*) is a vector lattice and $E^b = E^\#$.*

Proof.   Let $f \in E^b$. Define for each u in C that

$$g(u) = \sup\{f(x) : x \in [0, u]\}. \tag{1.2}$$

Clearly g is positively homogeneous. Also, if $u, v \in C$ then

$$
\begin{aligned}
g(u) + g(v) &= \sup\{f(x) : x \in [0, u]\} + \sup\{f(y) : y \in [0, v]\} \\
&= \sup\{f(x) + f(y) : x \in [0, u], y \in [0, v]\} \\
&= \sup\{f(x+y) : x \in [0, u], y \in [0, v]\} \\
&= \sup\{f(z) : z \in [0, u+v]\},
\end{aligned}
$$

where the last equality holds since E has the Riesz decomposition property. Therefore g is additive on C. If $x = u - v \in E = C - C$, where $u, v \in C$, then we define $g(x) = g(u) - g(v)$. It is easy to verify that g is a well-defined linear functional on E and agrees with the formula (1.2) on C. It is easy to see that $g \in E^*$. By formula (1.2), $0, f \leqslant g$ so $g \in C^* \subseteq E^b$; in fact it is not difficult to verify that $g = \sup\{0, f\}$. Consequently, E^b is a vector lattice. Note also that $f = g - (g - f) \in C^* - C^* = E^\#$. This shows that $E^b \subseteq E^\#$; hence $E^b = E^\#$.

(1.11) PROPOSITION.   *Let (E, C) be an ordered vector space, and let $f \in E^b$. Then for each u in C we have*

$$\sup f[0, u] + \inf f[0, u] = f(u). \tag{1.3}$$

Proof.   If $u = 0$, the equality (1.3) is trivial. We may therefore suppose that $u \in C$ and $u \neq 0$. Let $y \in [0, u]$. Then $u - y \in [0, u]$ and

$$\inf f[0, u] \leqslant f(u - y) = f(u) - f(y) \leqslant \sup f[0, u],$$

hence

$$f(y) + \inf f[0, u] \leqslant f(u) \leqslant f(y) + \sup f[0, u], \quad \text{for all } y \in [0, u].$$

The equality (1.3) follows easily.

We shall now turn our attention to a problem concerned with positive extensions of linear functionals. The following theorem is due, independently, to Namioka (1957) and Bauer (1957).

(1.12) THEOREM. *Let F be a vector subspace of an ordered vector space* (E, C), *and let f be a linear functional on F. Then the following statements are equivalent:*

(a) *f can be extended to a positive linear functional on E;*

(b) *there exists a convex and absorbing subset V of E such that*

$$f(x) \leqslant 1 \quad \text{for all } x \in F \cap (V - C).$$

Proof. The implication (a) $\Rightarrow$ (b) is clear; in fact, if g is a positive extension of f on E, then the set

$$V = \{x \in E : g(x) \leqslant 1\}$$

has the desired property. To prove the implication (b) $\Rightarrow$ (a), we note that $V - C$ is convex and absorbing because of $V \subseteq V - C$. Suppose that p is the gauge of $V - C$; then p is a sublinear functional on E and

$$f(x) \leqslant p(x) \qquad \text{for all } x \in F.$$

By making use of the Hahn–Banach theorem, there exists a linear functional g on E which is an extension of f and

$$g(y) \leqslant p(y) \quad \text{for all } y \in E.$$

Note that

$$-C \subseteq V - C \subseteq \{x \in E : p(x) \leqslant 1\}$$

and that C is a cone; we conclude that g must be positive.

Let (E, C) be an ordered vector space. A subspace F of E is said to be *cofinal* if $C \subseteq F - C$.

(1.13) COROLLARY. *Let F be a cofinal subspace of an ordered vector space* (E, C). *Then every positive linear functional on F can be extended to a positive linear functional on E.*

Proof. Without loss of generality we can assume that C is generating. Let f be a positive linear functional on F, and let

$$U = \{x \in F : f(x) \leqslant 1\}.$$

Suppose that

$$V = U - C.$$

Then it is not hard to verify that V is convex and absorbing because F is cofinal. Since C is a cone, it follows that

$$F \cap (V-C) = F \cap (U-C)$$

and that

$$f(x) \leqslant 1 \qquad \text{for all } x \in F \cap (V-C).$$

The result now follows from the preceding theorem.

An element e in an ordered vector space (E, C) is called an *order-unit* if the order-interval $[-e, e]$ is absorbing in E, that is, if $e \in C$ and for each x in E there exists $\lambda > 0$ such that $-\lambda e < x < \lambda e$. More generally, a net $\{e_\lambda, \lambda \in \Lambda, \leqslant\}$ in E is called an *approximate order-unit* if the following conditions are satisfied:

(a) each e_λ is in C;

(b) for any pair of elements λ_1, λ_2 in the directed set Λ with $\lambda_1 \leqslant \lambda_2$, it is true that $e_{\lambda_1} \leqslant e_{\lambda_2}$;

(c) for each x in E there exist $\lambda \in \Lambda$ and a positive real number α such that $-\alpha e_\lambda \leqslant x \leqslant \alpha e_\lambda$.

Thus, if $\{e_\lambda\}$ is an approximate order-unit then the set

$$S_\Lambda = \cup [-e_\lambda, e_\lambda] = \{x \in E : -e_\lambda \leqslant x \leqslant e_\lambda \quad \text{for some } \lambda \in \Lambda\}$$

is circled, convex, and absorbing in E. Thus the Minkowski functional of S_Λ is a semi-norm on E and will be referred to as the *approximate order-unit semi-norm* defined by $\{e_\lambda\}$. If e is an order-unit in E then the Minkowski functional of $[-e, e]$ is called an *order-unit semi-norm* defined by e.

If e is an order-unit in (E, C) then the vector ordering is Archimedean if and only if $x \leqslant 0$ whenever $x \leqslant \alpha e$ for all $\alpha > 0$ and is almost-Archimedean if and only if $x = 0$ whenever $-\alpha e \leqslant x \leqslant \alpha e$ for all $\alpha > 0$.

(1.14) COROLLARY. *Let (E, C) be an ordered vector space with an order-unit (or, more generally, an approximate order-unit $\{e_\lambda\}$), and let F be a vector subspace of E containing e (or $\{e_\lambda\}$). Then each positive linear functional on F can be extended to a positive linear functional on E.*

Let (E, C) be an ordered vector space. A functional q, defined on C, is said to be *superlinear* if $-q$ is sublinear, i.e.

$$q(\lambda u) = \lambda q(u) \quad \text{for all } \lambda \geqslant 0, u \in C, \quad \text{and}$$

$$q(u+\omega) \geqslant q(u)+q(\omega) \quad \text{for all } u, \omega \text{ in } C.$$

The following generalization of the Hahn–Banach theorem was proved by Bonsall (1955) and will be very useful in our subsequent discussions.

(1.15) THEOREM (Bonsall). *Let (E, C) be an ordered vector space and p a sublinear functional on E, and suppose that q is a superlinear functional on C such that*

$$q(u) \leqslant p(u) \quad \text{for all } u \text{ in } C.$$

Then there exists a linear functional f on E such that

$$f(x) \leqslant p(x) \quad \text{for all } x \in E,$$

and

$$q(u) \leqslant f(u) \quad \text{for all } u \in C.$$

Proof. Define, for every $x \in E$, that

$$r(x) = \inf\{p(x+u) - q(u) : u \in C\}.$$

Since $p(u) \leqslant p(x+u) + p(-x)$, it follows that $r(x) \geqslant -p(-x)$, and hence that r is finite on E. Clearly r is a sublinear functional on E, $r(x) \leqslant p(x)$ for all x in E, and $r(-u) \leqslant -q(u)$ for all u in C. By the Hahn–Banach theorem, there exists a linear functional f on E such that $f(x) \leqslant r(x)$ for all $x \in E$, and so $f(x) \leqslant p(x)$ for all $x \in E$. Since

$$f(-u) \leqslant r(-u) \leqslant -q(u) \quad \text{for all } u \in C,$$

we conclude that $f(u) \geqslant q(u)$ for all $u \in C$. This completes the proof.

Bonsall's theorem will be useful in our investigation of the duality problems. To facilitate its applications, we introduce the following notation. For any subset U of an ordered vector space (E, C), we define

$$S(U) = \cup \{[-u, u] : u \in U \cap C\}.$$

Then U is absolutely dominated if and only if $S(U) \supseteq U$; U is absolute order-convex if and only if $S(U) \subseteq U$, and U is solid if and only if $S(U) = U$.

If $A \subseteq E$ the *polar* of A, taken in the algebraic dual E^*, will be denoted by A^π and defined by

$$A^\pi = \{f \in E^* : f(a) \leqslant 1, \forall a \in A\}.$$

For example, $C^\pi = -C^*$.

(1.16) LEMMA. *Let V be a subset of an ordered vector space (E, C). Then we have*

$$S(V^\pi) \subseteq (S(V))^\pi.$$

Consequently, if V is absolutely dominated then the polar V^{π} of V, taken in E^*, is absolute order-convex in E^*.

Proof. Let f be in $S(V^{\pi})$. Then there is $0 < g \in V^{\pi}$ such that $\pm f < g$. Let $x \in S(V)$ and suppose that $\pm x < v$ for some $v \in V$. Then

$$0 < (g-f)(v+x) = g(v)+g(x)-f(v)-f(x)$$

and

$$0 < (g+f)(v-x) = g(v)-g(x)+f(v)-f(x).$$

Summing up, it follows that

$$0 < 2g(v)-2f(x).$$

Since $v \in V$ and $g \in V^{\pi}$, we then have $f(x) < g(v) < 1$, valid for all x in $S(V)$. This shows that $f \in (S(V))^{\pi}$ and hence that $S(V^{\pi}) \subseteq (S(V))^{\pi}$. Further, if V is absolutely dominated, then $V \subseteq S(V)$; hence

$$S(V^{\pi}) \subseteq (S(V))^{\pi} \subseteq V^{\pi};$$

that is, V^{π} is absolute-order-convex.

(1.17) THEOREM (Jameson).   *Let (E, C) be an ordered vector space, and let V be a convex and absorbing subset of E. Then*

$$(S(V))^{\pi} = S(V^{\pi}).$$

Consequently, if V is absolute-order-convex then V^{π} is absolutely dominated, and if V is solid then V^{π} is solid.

Proof. In view of the preceding lemma, we have only to show that $(S(V))^{\pi} \subseteq S(V^{\pi})$. Let g be in $(S(V))^{\pi}$. We have to find an f with $0 < f \in V^{\pi}$ such that $\pm g < f$. To do this, we define for each $x \in C$ that

$$q(x) = \sup\{g(y):y \in E, \pm y < x\}.$$

Since $g \in (S(V))^{\pi}$ and since V is absorbing, it is easy to see that q is a well-defined superlinear functional on C. Further, let p be the gauge of V. Notice that

$$\pm y < x \in V \Rightarrow g(y) < 1$$

and hence that

$$q(v) < p(v) \quad \text{for all } v \in C.$$

By theorem (1.15), there exists $f \in E^*$ such that

$$f(y) < p(y) \quad \text{for all } y \in E,$$

and

$$q(v) < f(v) \quad \text{for all } v \in C.$$

Notice that $\pm g(v) \leqslant q(v)$ for all $v \in C$; hence $\pm g(v) \leqslant f(v)$ for all $v \in C$ and $\pm g \leqslant f$. Notice also that $f(y) \leqslant p(y) \leqslant 1$ for all $y \in V$ so that $f \in V^\pi$. This completes the proof for the first assertion of the theorem. Furthermore, if V is absolute-order-convex then $S(V) \subseteq V$ and, by the first part of the proof,

$$V^\pi \subseteq S(V)^\pi = S(V^\pi);$$

i.e. V^π is absolutely dominated. This, together with lemma (1.16), implies that if V is solid then so is V^π.

There do not always exist non-zero positive linear functionals on an arbitrary ordered vector space (cf. Jameson (1970), p. 2). By virtue of the preceding theorem we are now in a position to derive a condition which is necessary and sufficient for the existence of a non-zero positive linear functional on an ordered vector space E.

(1.18) PROPOSITION.   *Let (E, C) be an ordered vector space. Then the following statements are equivalent:*
  (a) *there exists a non-zero positive linear functional on E;*
  (b) *there exists a non-zero absolute-monotone semi-norm on E;*
  (c) *there exists a non-zero monotone semi-norm on E.*

Proof.   (a) $\Rightarrow$ (b): If f is a non-zero positive linear functional on E then the semi-norm p, defined by

$$p(x) = |f(x)| \qquad (x \in E),$$

is a non-zero absolute-monotone semi-norm on E.
  (b) $\Rightarrow$ (c): Trivial.
  (c) $\Rightarrow$ (a): Let p be a non-zero monotone semi-norm on E. Then there exists x_0 in E such that $p(x_0) \neq 0$. By the Hahn–Banach theorem, there exists a linear functional g on E such that $g(x_0) = p(x_0)$ and

$$g(x) \leqslant p(x) \quad (x \in E).$$

Notice that $g \in U^\pi$, where $U = \{x \in E : p(x) \leqslant 1\}$. Also, since p is monotone, $S(U) \subseteq 3U$; in fact, if $-x \leqslant y \leqslant x$ and $x \in U$, then $0 \leqslant x+y \leqslant 2x$; hence

$$p(y) \leqslant p(x+y)+p(-x) \leqslant p(x+y)+p(x) \leqslant 2+1 = 3.$$

By the preceding theorem, $(S(U))^\pi = S(U^\pi)$, and it follows that

$$\tfrac{1}{3}U^\pi \subseteq (S(U))^\pi = S(U^\pi).$$

In particular, $g \in 3 \cdot S(U^{\pi})$; hence $\pm g \leqslant 3f$ for some $f \in U^{\pi}$ with $f > 0$. Let $g_1 = \frac{1}{2}(3f+g)$ and $g_2 = \frac{1}{2}(3f-g)$. Then g_1, g_2 are *positive* linear functionals on E and $g = g_1 - g_2$. Since $g(x_0) \neq 0$, one of g_1 and g_2 must be *non-zero* at x_0.

Suppose that (E, C) is an ordered vector space. A sequence $\{x_n : n \in \mathbf{N}\}$ in E is said to be *increasing*, and we write $x_n\!\uparrow$, if $x_n \leqslant x_m$ whenever $n < m$. The sequence is said to be *decreasing*, and we write $x_n\!\downarrow$, if $x_m \leqslant x_n$ whenever $n < m$. If $x_n\!\uparrow$ and $x = \sup x_n$ exists in E, we then write $x_n\!\uparrow x$. Similarly for a decreasing sequence. The index set $\{x_\tau : \tau \in D\}$ (abbreviated to $\{x_\tau\}$) of E is said to be *directed upwards*, denoted by $x_\tau\!\uparrow$, if for every pair τ_1 and τ_2 in D there exists τ in D such that $x_{\tau_1} \leqslant x_\tau$ and $x_{\tau_2} \leqslant x_\tau$. It is said to be *directed downwards*, denoted by $x_\tau\!\downarrow$, if for every pair τ_1 and τ_2 in D there exists τ in D such that $x_{\tau_1} \geqslant x_\tau$ and $x_{\tau_2} \geqslant x_\tau$. If $\{x_\tau : \tau \in D\}$ is a directed upwards subset of E, and we define a relation '$\leqslant$' in D by

$$\tau_1 \leqslant \tau_2 \quad \text{if} \quad x_{\tau_1} \leqslant x_{\tau_2},$$

then $(D, \leqslant)$ is a directed set, and hence $\{x_\tau : \tau \in D\}$ is a net with the property that $x_{\tau_1} \leqslant x_{\tau_2}$ whenever $\tau_1 \leqslant \tau_2$; in this case, $\{x_\tau : \tau \in D\}$ is called an *increasing net*. If $\{x_\tau : \tau \in D\}$ is a directed downwards subset of E, and we define a relation '$\leqslant$' in D by $\tau_1 \leqslant \tau_2$ if $x_{\tau_1} \geqslant x_{\tau_2}$, then $(D, \leqslant)$ is a directed set, and hence $\{x_\tau : \tau \in D\}$ is a net with the property that $x_{\tau_1} \geqslant x_{\tau_2}$ whenever $\tau_1 \leqslant \tau_2$; in this case, $\{x_\tau : \tau \in D\}$ is called a *decreasing net*. If $x_\tau\!\uparrow$ and if $x = \sup x_\tau$ exists in E, we write $x_\tau\!\uparrow x$. Similarly for sets directed downwards. A subset B of E is said to be *order-complete* if every increasing net in B that is majorized in E has a supremum which belongs to B, and is *σ-order-complete* if every increasing sequence in B that is majorized in E has a supremum in B. In particular, if E itself is order-complete, we then say that E is an *order-complete* vector space or if E is *σ-order-complete*, then E is an *σ-order-complete* vector space. A subspace of E which is order-convex is called an o-*ideal*. A solid subspace B of E is called a *normal subspace* of E if it follows from $x_\tau \uparrow x$ in E with x_τ in B for all τ that x belongs to B; it is called a *σ-normal subspace* if it follows from $x_n \uparrow x$ in E with x_n in B for all n that x belongs to B. It is clear that each solid subspace of E which is order-complete must be a normal subspace of E, and that if E is an order-complete vector space, then each normal subspace of E is an order-complete subset of E.

If $\{(E_\alpha, C_\alpha) : \alpha \in I\}$ is a family of ordered vector spaces then $\prod_{\alpha \in I} C_\alpha$ is a cone in the product space $\prod_{\alpha \in I} E_\alpha$. Notice that $\prod_{\alpha \in I} C_\alpha$ is a proper cone if and only if each C_α is proper. If (E, C) is an ordered vector space and if E is the algebraic direct sum of subspaces F_i $(i = 1, 2,..., n)$ of E, we say that E is the *ordered direct sum* of F_i $(i = 1, 2,..., n)$ if $C = \prod_{i=1}^{n} C_i$, where $C_i = C \cap F_i$ $(i = 1, 2,..., n)$. It is easily seen that C is a proper cone if and only if each C_i is a proper cone in F_i, that C is generating if and only if each C_i is generating, and that the vector ordering determined by C is Archimedean if and only if each vector ordering determined by C_i is Archimedean.

CONES IN TOPOLOGICAL VECTOR SPACES

IT is easily seen that the closure of a cone in a topological vector space E is a cone. But the closure of a proper cone in E need not be proper. By way of example, consider $\mathbf{R}^2$ with a cone C_l defined by

$$C_l = \{(x, y):x > 0\} \cup \{(0, y):y \geqslant 0\},$$

then C_l is a proper cone in $\mathbf{R}^2$, and the closure $\bar{C}_l$ of C_l with respect to the usual topology is

$$\{(x, y):x \geqslant 0\},$$

but $\bar{C}_l$ is not proper.

A topological vector space with a cone is called an *ordered topological vector space* and a locally convex space with a cone is referred to as an *ordered convex space*. It should be noted that in an ordered topological vector space the vector ordering and the vector topology need not have any connection. However, if there exists a closed cone C in a topological vector space $(E, \mathscr{P})$, then the vector topology $\mathscr{P}$ must be Hausdorff because $C \cap -C = \{0\}$ is $\mathscr{P}$-closed; but the converse is, in general, not true; for instance, consider $\mathbf{R}^2$ with a cone C_0 defined by

$$C_0 = \{(x, y):x \geqslant 0 \quad \text{and} \quad y > 0\} \cup \{(0, 0)\},$$

then C_0 is a cone in $\mathbf{R}^2$, but it is not closed with respect to the usual topology.

In what follows, all topological vector spaces will be assumed to be Hausdorff, unless a statement is made to the contrary.

If $(E, C, \mathscr{P})$ is an ordered convex space, throughout this book E' will denote the topological dual of E, E^{tb} will denote the set of all $\mathscr{P}$-bounded linear functionals on E, and C' will denote the cone consisting of all positive $\mathscr{P}$-continuous linear functionals on E, that is, $C' = E' \cap C^*$. E^{tb} is referred to as the *topologically bounded dual of* E.

(2.1) PROPOSITION. *Let $(E, C, \mathscr{P})$ be an ordered topological vector space and let $\leqslant$ be the vector ordering determined by C. Then we have:*

(a) *the vector ordering $\leqslant$ is Archimedean if C is $\mathscr{P}$-closed;*

(b) *the vector ordering $\leqslant$ is almost-Archimedean if the $\mathscr{P}$-closure $\bar{C}$ of C is proper;*

(c) *if C is $\mathscr{P}$-closed and if $\{x_\tau : \tau \in D\}$ is an increasing net in E which converges to x with respect to $\mathscr{P}$, then $x = \sup x_\tau$; if C is $\mathscr{P}$-closed and if $\{x_\tau : \tau \in D\}$ is a decreasing net in E which converges to x with respect to $\mathscr{P}$, then $x = \inf x_\tau$.*

Proof. (a) Suppose that $nx \leqslant y$ for all positive integers n; we wish to show that $x \leqslant 0$. Since $\dfrac{1}{n}y \to 0$ with respect to $\mathscr{P}$, and $\dfrac{1}{n}y - x \in C$, we conclude from the closedness assumption that

$$-x = \lim_n \left(\frac{1}{n}y - x\right) \in C,$$

and hence that $x \leqslant 0$.

(b) Suppose that $-y \leqslant nx \leqslant y$ for all positive integers n. Then

$$-x = \lim_n \left(\frac{1}{n}y - x\right), \qquad x = \lim_n \left(\frac{1}{n}y + x\right).$$

On the other hand, since $\dfrac{1}{n}y - x \in C$ and since $\dfrac{1}{n}y + x \in C$, it follows that $-x \in \bar{C}$, $x \in \bar{C}$, and hence that $x \in \bar{C} \cap -\bar{C}$. We conclude, from the fact that $\bar{C}$ is proper, that $x = 0$; therefore the vector ordering $\leqslant$ is almost-Archimedean.

(c) Suppose that $x_\tau \uparrow$. We first show that $x_\tau \leqslant x$ for all $\tau \in D$. For any τ in D, let
$$A_\tau = \{x_\nu : \nu \geqslant \tau, \, \nu \in D\}.$$

Then $A_\tau \subseteq x_\tau + C$, and so the $\mathscr{P}$-closure $\bar{A}_\tau$ of A_τ is contained in $x_\tau + C$ because $x_\tau + C$ is $\mathscr{P}$-closed. Since x_τ converges to x, it follows that $x \in \bar{A}_\tau$, and hence that $x \in x_\tau + C$ or, equivalently, $x - x_\tau \geqslant 0$ for all τ in D. This shows that x is an upper bound of $\{x_\tau : \tau \in D\}$. If y in E is such that $x_\tau \leqslant y$ for all $\tau \in D$, then $y - x_\tau \in C$ for all $\tau \in D$. Since C is $\mathscr{P}$-closed and since $y - x_\tau$ converges to $y - x$, we conclude that $y - x \in C$, and hence that $x \leqslant y$. Therefore $x = \sup x_\tau$.

Finally if $x_\tau \downarrow$ then $\{-x_\tau : \tau \in D\}$ is directed upwards and $-x_\tau$ converges to $-x$ with respect to $\mathscr{P}$; therefore $-x = \sup(-x_\tau) = -\inf x_\tau$, consequently $x = \inf x_\tau$. This completes the proof.

Remark. It is easily seen that if $(E, C, \mathscr{P})$ is an ordered topological vector space and if C is $\mathscr{P}$-closed, then each order-interval in E is $\mathscr{P}$-closed. Furthermore, C is $\mathscr{P}$-closed if and only if the vector ordering $\leqslant$ induced by C is 'continuous' in the following sense: Whenever

$\{x_n\}$, $\{y_m\}$ are two convergent nets in $(X, \mathscr{P})$ with limits x and y such that $x_n \leqslant y_m$ for all m, n, then it is true that $x \leqslant y$.

(2.2) PROPOSITION. *Let $(E, C, \mathscr{P})$ be an ordered topological vector space. Then e is an interior point of C if and only if $[-e, e]$ is a $\mathscr{P}$-neighbourhood of 0; in this case, e is an order-unit element.*

Proof. (a) *Necessity.* Let e be an interior point of C. There exists a circled $\mathscr{P}$-neighbourhood V of 0 such that $e+V \subseteq C$; it follows that $V \subseteq (C-e) \cap (e-C)$ since V is circled. It is clear that

$$[-e,e] = (C-e) \cap (e-C),$$

consequently $[-e, e]$ is a $\mathscr{P}$-neighbourhood of 0.

(b) *Sufficiency.* Suppose that $[-e, e]$ is a $\mathscr{P}$-neighbourhood of 0. We conclude from $e+[-e, e] \subseteq C$ that e is an interior point of C.

Finally, e is an order-unit element because $[-e, e]$ is absorbing.

(2.3) COROLLARY. *Let $(E, C, \mathscr{P})$ be an ordered topological vector space. If the interior of C is non-empty, then $E^{\mathrm{b}} \subseteq E'$.*

Proof. Let e be an interior point of C, and let f be in E^{b}. By the preceding result, $[-e, e]$ is a $\mathscr{P}$-neighbourhood of 0. Since f is bounded on $[-e, e]$, there exists $\lambda > 0$ such that

$$f(y) \leqslant \lambda \qquad \text{for all } y \in [-e, e].$$

It then follows that $[-e, e] \subseteq \{y \in E : f(y) \leqslant \lambda\}$, and hence that $f \in E'$.

(2.4) COROLLARY. *Let $(E, C, \mathscr{P})$ be an ordered topological vector space, and let the interior of C be non-empty. If e is an order-unit element, then e is an interior point of C.*

Proof. Let x be an interior point of C. By proposition (2.2), $[-x, x]$ is a $\mathscr{P}$-neighbourhood of 0. Since e is an order-unit element in E, there exists $\lambda > 0$ such that

$$-\lambda e \leqslant x \leqslant \lambda e.$$

Clearly $e+\lambda^{-1}[-x, x] \subseteq C$, and so e is an interior point of C.

For a barrelled space with a closed cone, the condition that the interior of C is non-empty in the preceding result can be dropped.

(2.5) COROLLARY. *Let $(E, C, \mathscr{P})$ be an ordered convex space, and let C be $\mathscr{P}$-closed. If $(E, \mathscr{P})$ is barrelled, then each order-unit element is an interior point of C.*

Proof. Let e be an order-unit element. Since C is $\mathscr{P}$-closed, $[-e, e]$ is $\mathscr{P}$-closed, and so $[-e, e]$ is a barrel, consequently $[-e, e]$ is a $\mathscr{P}$-neighbourhood of 0. By proposition (2.2), e is an interior point of C. This completes the proof.

(2.6) PROPOSITION. *Let $(E, C, \mathscr{P})$ be an ordered convex space and let E^{tb} be the set of all $\mathscr{P}$-bounded linear functionals on E. Then $E^{\mathrm{tb}} \subseteq E^{\mathrm{b}}$ if and only if each order-bounded subset of E is $\mathscr{P}$-bounded.*

Proof. The sufficiency is clear. We use the Mackey–Arens theorem (cf. Schaefer (1966)) to verify the necessity. Let B be an order-bounded subset of E, and $f \in E'$. Then $f \in E^{\mathrm{tb}} \subseteq E^{\mathrm{b}}$. Hence f is bounded on B. This shows that B is $\sigma(E, E')$-bounded, hence $\mathscr{P}$-bounded.

In theorem (1.12) we considered a positive linear extension problem. The following result deals with *continuous* and positive linear extensions; the proof is exactly that given for theorem (1.12).

(2.7) PROPOSITION. *Let F be a vector subspace of an ordered convex space $(E, C, \mathscr{P})$. Then a linear functional f defined on F has a $\mathscr{P}$-continuous, positive linear extension to E if and only if there exists a convex $\mathscr{P}$-neighbourhood V of 0 such that*

$$f(x) \leqslant 1 \qquad \text{for all } x \in F \cap (V - C).$$

(2.8) PROPOSITION. *Let $(E, C, \mathscr{P})$ be an ordered topological vector space with the topological dual E' and let $f \in E'$. Then the following statements are equivalent:*

(a) *$f \in C' - C'$;*

(b) *there exists a convex $\mathscr{P}$-neighbourhood V of 0 and a positive constant α such that $f(x) \leqslant \alpha$ for all $x \in C \cap (V - C)$;*

(c) *there exists an o-convex circled $\mathscr{P}$-neighbourhood W of 0 and a positive constant β such that $f(x) \leqslant \beta$ for all $x \in W$.*

Proof. (a) $\Rightarrow$ (b): Suppose that $f = g - h$ where g, h are in C'. Let $V = \{x \in E : g(x) \leqslant 1\}$. Then V is a convex $\mathscr{P}$-neighbourhood of 0. Further, if $x \in C \cap (V - C)$ then $0 \leqslant x \leqslant v$ for some v in V. Hence

$$f(x) = g(x) - h(x) \leqslant g(x) \leqslant g(v) \leqslant 1.$$

(b) $\Rightarrow$ (c): We can further assume that V is symmetric. Let $U = \{x \in V : |f(x)| \leqslant \alpha\}$ and W the order-convex hull of U, i.e. $W = (U+C) \cap (U-C)$. Since $f \in E'$, it follows from

$$U = V \cap \{x \in E : |f(x)| \leqslant \alpha\}$$

that U is a $\mathscr{P}$-neighbourhood of 0, consequently W is an o-convex $\mathscr{P}$-neighbourhood of 0. Let $w \in W$ and assume that $u \leqslant w \leqslant v$ for some $u, v \in U$. Then $|f(u)| \leqslant \alpha$. Also, since $0 \leqslant (w-u)/2 \leqslant (v-u)/2 \in V$, we have $(w-u)/2 \in C \cap (V-C)$ so that

$$f\!\left(\frac{w-u}{2}\right) \leqslant \alpha.$$

Consequently

$$f(w) = f(w-u)+f(u) \leqslant f(w-u)+|f(u)| \leqslant 3\alpha,$$

valid for all $w \in W$.

(c) $\Rightarrow$ (a): Since W is a neighbourhood, the polar W^0 of W, taken in E', is the same as that taken in E^*. Let $g = f/\beta$. Then $g \in W^0$. Since W is order-convex, $S(W) \subseteq W$; it follows from the duality theorem (1.17) that
$$g \in W^\pi \subseteq S(W)^\pi = S(W^\pi) = S(W^0).$$

Therefore there exists $h \in W^0$ such that $\pm g \leqslant h$. Then

$$g = \tfrac{1}{2}(h+g)-\tfrac{1}{2}(h-g) \in C'-C'.$$

This completes the proof of the proposition.

Next we shall give a duality theorem parallel to the duality theorem (1.17). We first prove a simple computation rule for polars. Recall that if A is a subset of a topological vector space E, the polar of A, taken in the topological dual E', is denoted by A^0; thus $A^0 = A^\pi \cap E'$.

(2.9) LEMMA. *Let S and T be two convex subsets of a locally convex space E containing the origin 0. Then the following propositions hold:*

(a) *if $\overline{S \cap T} = \overline{S} \cap \overline{T}$ (for example, S and T are closed) then $(S \cap T)^0 = \overline{\mathrm{co}}(S^0 \cup T^0)$, where the polars are taken in the topological dual E' and $\overline{\mathrm{co}}(S^0 \cup T^0)$ denotes the $\sigma(E', E)$-closed convex hull of $S^0 \cup T^0$;*

(b) *if 0 is an interior point of S and of T then $\overline{S \cap T} = \overline{S} \cap \overline{T}$.*
(Thus, roughly speaking, if S and T are both open or both closed convex subsets of E containing 0, then the polar of the intersection equals to the closed convex hull of polars.)

Proof. Notice that $\bar{S}$ and $\bar{T}$ are closed convex subsets of E containing 0. By the bipolar theorem, we have

$$(\bar{S} \cap \bar{T})^0 = \overline{\mathrm{co}}(\bar{S}^0 \cup \bar{T}^0).$$

Since $\bar{S} \cap \bar{T} = \overline{S \cap T}$ and the polar (in E') of a set is the same as the polar of its closure, we have

$$(S \cap T)^0 = (\overline{S \cap T})^0 = (\bar{S} \cap \bar{T})^0 = \overline{\mathrm{co}}(\bar{S}^0 \cup \bar{T}^0) = \overline{\mathrm{co}}(S^0 \cup T^0),$$

proving (a).

To prove (b), let $x \in \bar{S} \cap \bar{T}$. Since S is a convex set containing the origin as an interior point, it is easily seen that $\lambda x \in S$ for each $0 \leqslant \lambda < 1$ (λx is in fact an interior point of S, cf. Schaefer (1966, p. 38)). Similarly $\lambda x \in T$. Letting $\lambda \to 1$ in $\lambda x \in S \cap T$, we have $x \in \overline{S \cap T}$. This shows that $\bar{S} \cap \bar{T} \subseteq \overline{S \cap T}$ and consequently $\bar{S} \cap \bar{T} = \overline{S \cap T}$ since the opposite inclusion is obvious.

Remark. If we assume one of the sets S, T to be a neighbourhood of 0 and the other set is closed, then (b) remains true. If both sets are neighbourhoods of 0 then S^0, T^0 are $\sigma(E', E)$-compact by the Alaoglu theorem (cf. Schaefer (1966)); hence $\mathrm{co}(S^0 \cup T^0)$ is $\sigma(E', E)$-closed and $(S \cap T)^0 = \mathrm{co}(S^0 \cup T^0)$.

(2.10) COROLLARY. *Let S and T be two convex absorbing subsets of a vector space E containing the origin 0. Then, for polars taken in the algebraic dual E^*, we have*

$$(S \cap T)^\pi = \mathrm{co}(S^\pi \cup T^\pi).$$

Proof. Let τ denote the Mackey topology in E with respect to the duality $\langle E, E^* \rangle$. Then all convex absorbing sets in E are neighbourhoods of 0 in (E, τ). Hence the corollary follows from the remark preceding the statement of the corollary.

(2.11) THEOREM. *Let V be a circled, convex, absorbing subset of an ordered vector space (E, C). Then the following propositions hold:*
  (a) $(F(V))^\pi = D(V^\pi)$;
  (b) $(D(V))^\pi = F(V^\pi)$;
where $D(V^\pi)$ and $F(V^\pi)$ respectively denotes the decomposable kernel and the order-convex hull of V^π in the algebraic dual E^*.

Proof. (a) We first recall that $C^\pi = -C^*$. Next we show that

$$(V+C)^\pi = V^\pi \cap C^\pi = -(V^\pi \cap C^*). \tag{2.1}$$

In fact, since V, C contain 0, we have $V+C \supseteq V, C$; hence $(V+C)^\pi \subseteq V^\pi \cap C^\pi$. On the other hand, if $f \in V^\pi \cap C^\pi$ then $-f \in V^\pi \cap C^*$ (since V is circled); hence $(-f)(c) \geqslant 0$ for all $c \in C$. Consequently

$$f(v+c) = f(v)+f(c) \leqslant f(v) \leqslant 1 \qquad (v \in V, c \in C),$$

showing that $f \in (V+C)^\pi$. Therefore formula (2.1) holds. Similarly, we can show that $(V-C)^\pi = V^\pi \cap C^*$. On applying corollary (2.10), we then have

$$(F(V))^\pi = ((V+C) \cap (V-C))^\pi = \mathrm{co}((V+C)^\pi \cup (V-C)^\pi)$$
$$= \mathrm{co}(-(V^\pi \cap C^*) \cup (V^\pi \cap C^*)) = D(V^\pi).$$

(b) Let $f \in (D(V))^\pi$. Then $f(x) \leqslant p(x)$ for all $x \in C$, where p denotes the gauge of V. By Bonsall's generalization of the Hahn–Banach theorem (theorem (1.15)), there exists a linear functional g on E such that $f(x) \leqslant g(x)$ and $g(y) \leqslant p(y)$ for all $x \in C$ and $y \in E$. Then $g \in V^\pi$ and $f \leqslant g$. Similarly, considering $-f$ instead of f, we can find $h \in V^\pi$ such that $h \leqslant f$. Hence $f \in F(V^\pi)$. This shows that $(D(V))^\pi \subseteq F(V^\pi)$. Conversely, let $f' \in F(V^\pi)$, and suppose $x = \lambda_1 x_1 - \lambda_2 x_2$, $h' \leqslant f' \leqslant g'$, where λ_1, $\lambda_2 \geqslant 0$, $\lambda_1 + \lambda_2 = 1$, x_1, $x_2 \in V \cap C$, and h', $g' \in V^\pi$. To complete the proof we have to show that $f'(x) \leqslant 1$. To this end, we notice that $f'(x_1) \leqslant g'(x_1) \leqslant 1$ and $f'(-x_2) \leqslant h'(-x_2) \leqslant 1$ (since V is circled, $-x_2 \in V$). Hence

$$f'(x) = \lambda_1 f'(x_1) + \lambda_2 f'(-x_2) \leqslant \lambda_1 + \lambda_2 = 1.$$

The proof of the theorem is completed.

(2.12) CﻗOROLLARY (Jameson). *Let V be a circled, convex, absorbing subset of an ordered vector space (E, C), and suppose that V is order-convex. Let f be a linear functional on E such that $\sup f(V) < \infty$. Then there exist positive linear functionals g, h such that*

$$f = g-h \quad and \quad \sup g(V)+\sup h(V) = \sup f(V).$$

Proof. Without loss of generality, we can assume that $\sup f(V) = 1$. Then $f \in V^\pi = (F(V))^\pi$. By the preceding theorem, $f \in D(V^\pi)$; thus there exist positive linear functionals f_1, $f_2 \in V^\pi$ such that

$$f = \lambda_1 f_1 - \lambda_2 f_2$$

for some λ_1, $\lambda_2 \geqslant 0$ with $\lambda_1 + \lambda_2 = 1$. Let $g = \lambda_1 f_1$ and $h = \lambda_2 f_2$. Then $f = g - h$ and, for each v in the circled set V, it is true that

$$f(v) = g(v) - h(v) = g(v) + h(-v) \leqslant \sup g(V) + \sup h(V);$$

hence $1 = \sup f(V) \leqslant \sup g(V) + \sup h(V)$. On the other hand, since $g = \lambda_1 f_1$ and $f_1 \in V^\pi$, we know that $\sup g(V) \leqslant \lambda_1$. Similarly, $\sup h(V) \leqslant \lambda_2$. Consequently,

$$\sup g(V) + \sup h(V) \leqslant \lambda_1 + \lambda_2 = 1 \leqslant \sup f(V).$$

Therefore
$$\sup f(V) = \sup g(V) + \sup h(V).$$

(2.13) COROLLARY. *Let $(E, C, \mathscr{P})$ be an ordered convex space and V a circled convex $\mathscr{P}$-neighbourhood of 0 in E. If the topological dual E' is order-convex in E^* then $(D(V))^0 = F(V^0)$.*

Proof. Since E' is order-convex in E^*, the order-convex hull in E' of a subset A in E' is the same as the order-convex hull of A in E^*; i.e.
$$F(A) = (A + C') \cap (A - C') = (A + C^*) \cap (A - C^*).$$

By (b) of theorem (2.11),

$$(D(V))^\pi = F(V^\pi) = F(V^0)$$

and the above sets are contained in E'; thus the equalities can be rewritten as
$$(D(V))^0 = F(V^0).$$

The following result is dual to (b) of theorem (2.11).

(2.14) THEOREM. *Let $(E, C, \mathscr{P})$ be an ordered convex space with a $\mathscr{P}$-closed cone C, and let E' be the topological dual space with the dual cone C'. Let V be a circled convex $\mathscr{P}$-closed neighbourhood of 0 in E, and let $F(V^0)$ denote the order-convex hull of V^0 in (E', C'). Then*

$$(F(V^0))^0 = \overline{D(V)},$$

where the polars are taken with the duality $\langle E, E' \rangle$ and closure relative to $\mathscr{P}$.

(*Remark.* Since $D(V)$ is convex, the $\mathscr{P}$-closure $\overline{D(V)}$ of $D(V)$ is the $\sigma(E, E')$-closure of $D(V)$.)

Proof. As in formula (2.1) in the proof of theorem (2.11), we have

$$(V^0 + C')^0 = V^{00} \cap (C')^0 = -V^{00} \cap C^{00} = -V \cap C,$$

since V and C are closed. Also, by the Alaoglu–Bourbaki theorem, V^0 is $\sigma(E', E)$-compact, hence $V^0 + C'$ is $\sigma(E', E)$-closed (and convex). By lemma (2.9), we have

$$(F(V^0))^0 = ((V^0 + C') \cap (V^0 - C'))^0 = \overline{\text{co}}((V^0 + C')^0 \cup (V^0 - C')^0)$$

$$= \overline{\text{co}}(-(V \cap C) \cup (V \cap C) = \overline{D(V)}.$$

The following separation theorem will be useful in our further investigation.

(2.15) THEOREM. *Let $(E, C, \mathscr{P})$ be an ordered convex space with the topological dual E'. If K is a $\mathscr{P}$-compact convex subset of E, then there exists $f \in C'$ such that*

$$\sup f(K) < 0$$

if and only if $K \cap \bar{C} = \phi$, where $\bar{C}$ is the $\mathscr{P}$-closure of C.

Proof. (a) *Necessity.* If $K \cap \bar{C} \neq \phi$, take $x \in K \cap \bar{C}$; then there exists a net $\{x_\tau : \tau \in D\}$ in C such that x_τ converges to x with respect to $\mathscr{P}$, and so

$$f(x) = \lim_\tau f(x_\tau) \geqslant 0$$

because $f \in C'$; consequently

$$\sup f(K) \geqslant \sup\{f(x) : x \in K \cap \bar{C}\} \geqslant 0$$

which gives a contradiction.

(b) *Sufficiency.* Since $\bar{C}$ is $\mathscr{P}$-closed and convex, it follows from the strong separation theorem that there exists $f \in E'$ such that

$$\sup f(K) < \inf f(\bar{C}) = \inf f(C).$$

We claim that f is positive. In fact, if $u \in C$ then $nu \in C$ for all natural numbers n, and it follows from

$$\sup f(K) < f(nu) = nf(u) \qquad (n = 1, 2, \ldots)$$

that $f(u) \geqslant 0$; this shows that f is positive. Hence $\inf f(C) = 0$ and so $\sup f(K) < 0$. This completes the proof.

(2.16) COROLLARY. *Let $(E, C, \mathscr{P})$ be an ordered convex space, and let C be $\mathscr{P}$-closed. Then C is proper if and only if C' is total over E.*

Proof. By making use of theorem (2.15), we have

$$C = \{x \in E : f(x) \geqslant 0, \quad \forall f \in C'\},$$

and so
$$C \cap -C = \{x \in E \colon f(x) = 0, \ \forall f \in C'\};$$

consequently the result follows immediately.

We are now in a position to give a similar result to the Bonsall theorem (theorem (1.15)) with topological properties involved.

(2.17) PROPOSITION. *Let* $(E, C, \mathscr{P})$ *be an ordered convex space with a closed cone* C. *Let* K *be a convex compact subset of* E *containing the origin and let* F *be the cone generated by* K, *that is,* $F = \mathrm{pos}\, K$. *Let* p *be a sublinear functional on* F *and let* q *be a superlinear functional on* C *with the following properties:*

(a) q *is upper semi-continuous on* C;

(b) p *is lower semi-continuous on* K *and* $s = \sup |p(K)| < +\infty$;

(c) $q(x) < p(x)$ *for all* $x \in F \cap C$.

Then, for each $\varepsilon > 0$, *there exists* $f \in E'$ *such that*
$$q(x) \leqslant f(x) \quad and \quad f(y) < p(y) + \varepsilon$$

for all $x \in C$ *and* $y \in K$.

Proof. Consider the space $E \times \mathbf{R}$ with the product topology. Let
$$H = \{(a, \xi) \in E \times \mathbf{R} \colon a \in K, \ p(a) \leqslant \xi - \varepsilon \leqslant s\}$$
and
$$D = \{(a, \xi) \in E \times \mathbf{R} \colon a \in C, \ \xi \leqslant q(a)\}.$$

Then H is a compact, convex subset of $E \times \mathbf{R}$ containing $(a, p(a) + \varepsilon)$ for each a in K (in particular, containing $(0, \varepsilon)$), and D is a closed cone in $E \times \mathbf{R}$. Further, by (c), H and D are disjoint. Theorem (2.15) shows that there exists a continuous linear functional ψ on $E \times \mathbf{R}$ such that
$$0 = \sup \psi(D) < \inf \psi(H).$$

Since $(0, \varepsilon) \in H$, $0 < \psi(0, \varepsilon) = \varepsilon \cdot \psi(0, 1)$ so $\psi(0, 1) > 0$. Define μ on E by the rule
$$\mu(a) = -\frac{\psi(a, 0)}{\psi(0, 1)} \qquad (a \in E).$$

The μ is a continuous linear functional on E such that
$$\psi(a, \mu(a)) = 0 \qquad (a \in E).$$

If $y \in K$ then $(y, p(y) + \varepsilon) \in H$ and so
$$\psi(y, \mu(y)) < \psi(y, p(y) + \varepsilon);$$
since $\psi(0, 1) > 0$, it follows that $\mu(y) < p(y) + \varepsilon$ for all $y \in K$. Similarly,

since $(x, q(x)) \in D$, we can show that $q(x) \leqslant \mu(x)$ for all $x \in C$. Therefore μ is a required functional f, and the proof of the proposition is complete.

If $(E, C, \mathscr{P})$ is an ordered convex space, then C' is a $\sigma(E', E)$-closed cone in E'. Recall also that the topological dual of $(E', \sigma(E', E))$ is E. Thus we arrive at the following proposition.

(2.18) PROPOSITION. *Let $(E, C, \mathscr{P})$ be an ordered convex space. Let K be a convex $\sigma(E', E)$-compact subset of E' containing the origin, and let F be the cone generated by K. Let p^* be a sublinear functional on F and let q^* be a superlinear functional on C' with the following properties:*

(a) *q^* is upper semi-continuous on C' with respect to the relative $\sigma(E', E)$-topology;*

(b) *p^* is lower $\sigma(E', E)$-semi-continuous on K and*

$$s = \sup |p^*(K)| < +\infty$$

(c) *$q^*(f) \leqslant p^*(f)$ for all $f \in F \cap C'$.*
Then, for each $\varepsilon > 0$ there exists a $\sigma(E', E)$-continuous linear functional (and hence an element x in E) such that

$$q^*(f) \leqslant f(x) \quad and \quad g(x) \leqslant p^*(g) + \varepsilon$$

for all $f \in C'$ and $g \in K$.

Remark. Let $(E, \mathscr{P})$ be a locally convex space. Recall (cf. Baker (1968)) that a subset A of E' is said to be *almost $\sigma(E', E)$-closed* if its intersection with each $\sigma(E', E)$-closed equicontinuous subset of E' is $\sigma(E', E)$-closed. This is the case if and only if $A \cap V^0$ is $\sigma(E', E)$-closed for each $\mathscr{P}$-neighbourhood V of 0 in E. Thus any $\sigma(E', E)$-closed set is certainly almost $\sigma(E', E)$-closed. Following Baker (1968), a locally convex space $(E, \mathscr{P})$ is called a *hypercomplete space* if each *convex*, almost $\sigma(E', E)$-closed subset of E' is $\sigma(E', E)$-closed. In such a space, the condition (a) in the preceding proposition may be replaced by the following equivalent condition:

(a') q^* is upper semi-continuous on each equicontinuous subset of C' with respect to the relative $\sigma(E', E)$-topology.

In fact, (a) certainly implies (a'). Conversely, if q^* satisfies (a') and if λ is a real number, then let

$$A_\lambda = \{f \in C' : \lambda \leqslant q^*(f)\}.$$

Since q^* is superlinear on C', A_λ is convex. Further, if B is a $\sigma(E', E)$-closed and equicontinuous subset of C' then, by (a'), $q^*|_B$ is upper semi-continuous and hence

$$A_\lambda \cap B = \{f : \lambda \leqslant (q^*|_B)(f)\}$$

is a $\sigma(E', E)$-closed subset of B. Therefore $A_\lambda \cap B$ is $\sigma(E', E)$-closed in E', whenever B is $\sigma(E', E)$-closed and equicontinuous. Since E is hypercomplete, A_λ is $\sigma(E', E)$-closed and so q^* is upper semi-continuous on C'. This shows that (a') $\Rightarrow$ (a). Therefore in proposition (2.18), (a) may be replaced by (a') provided that E is hypercomplete. By the Krein–Smulian theorem (cf. Schaefer (1966)), any Banach space is hypercomplete. Thus we arrive at the following corollary (by an *ordered Banach space* we mean a Banach space with a partial ordering induced by a cone).

(2.19) COROLLARY. *Let $(E, C, \|.\|)$ be an ordered Banach space. Let K be a convex $\sigma(E', E)$-compact subset of E' containing the origin, and let F be the cone generated by K. Let p^* be a sublinear functional on F and let q^* be a superlinear functional on C' with the following properties:*

(a') *q^* is upper semi-continuous on $C' \cap \Sigma'$ with respect to the relative $\sigma(E', E)$-topology, where Σ' denotes the closed unit ball in $(E', \|.\|)$;*

(b) *p^* is lower $\sigma(E', E)$-semi-continuous on K and*

$$s = \sup |p^*(K)| < +\infty$$

(c) *$q^*(f) \leqslant p^*(f)$ for all $f \in F \cap C'$.*
Then for each $\varepsilon > 0$, there exists x in E such that

$$q^*(f) \leqslant f(x) \quad and \quad g(x) \leqslant p^*(g) + \varepsilon$$

for all $f \in C'$ and all $g \in K$.

Remark. In most applications, K is taken to be the closed unit ball Σ' or $\Sigma' \cap C'$. If $K = \Sigma'$ then the property

$$g(x) \leqslant p^*(g) + \varepsilon \qquad (g \in K)$$

implies that
$$\|x\| = \sup_{g \in \Sigma'} g(x) \leqslant \sup_{g \in \Sigma'} p^*(g) + \varepsilon = s + \varepsilon.$$

3

LOCALLY DECOMPOSABLE SPACES

In this chapter we study in detail the class of ordered topological vector spaces $(E, C, \mathscr{P})$ satisfying the property that each $V \cap C - V \cap C$ is a $\mathscr{P}$-neighbourhood of 0 whenever V is a $\mathscr{P}$-neighbourhood of 0. Such a property is called an *open decomposition property* and, in this case, we shall say that C gives an *open decomposition* in $(E, \mathscr{P})$. Recall that a subset B of an ordered topological vector space is positively generated if $B \subseteq B \cap C - B \cap C$. Thus C gives an open decomposition in $(E, \mathscr{P})$ if and only if $(E, \mathscr{P})$ admits a neighbourhood-base at 0 consisting of positively generated $\mathscr{P}$-neighbourhoods.

An ordered topological vector space $(E, C, \mathscr{P})$ is said to have a *nearly-open decomposition property* if $\overline{V \cap C - V \cap C}$ is a $\mathscr{P}$-neighbourhood of 0 whenever V is a $\mathscr{P}$-neighbourhood of 0. In such a case C is said to give a *nearly-open decomposition in* $(E, \mathscr{P})$.

Recall that if V is a subset of an ordered vector space (E, C) then $D(V)$ denotes the decomposable kernel of V. If V is circled convex, we have the following inequalities:

$$\tfrac{1}{2}(V \cap C - V \cap C) \subseteq D(V) \subseteq V \cap C - V \cap C;$$

thus $V \cap C - V \cap C$ is a $\mathscr{P}$-neighbourhood of 0 if and only if $D(V)$ is a $\mathscr{P}$-neighbourhood of 0.

A semi-norm p on (E, C) is said to be *semi-decomposable* if there exists a positive constant M (depending, in general, on p) such that the following condition holds: for any $x \in E$ and any $\varepsilon > 0$ there exist x_1, x_2 in C such that

$$x = x_1 - x_2 \quad \text{and} \quad p(x_1) + p(x_2) < M(p(x) + \varepsilon).$$

By a similar argument to that of proposition (1.7), p is semi-decomposable if and only if $V \subseteq MD(V)$, where $V = \{x \in E : p(x) < 1\}$. Thus the equivalence (a) $\Leftrightarrow$ (d) in the following theorem is clear.

(3.1) THEOREM. *Let $(E, C, \mathscr{P})$ be an ordered convex space. Then the following statements are equivalent:*

(a) *C gives an open decomposition in $(E, \mathscr{P})$;*

(b) $\mathscr{P}$ *admits a neighbourhood-base at* 0 *consisting of circled convex and decomposable sets;*

(c) $\mathscr{P}$ *admits a neighbourhood-base at* 0 *consisting of circled convex and absolutely dominated sets;*

(d) $\mathscr{P}$ *is defined by a family of semi-decomposable semi-norms* $\{p_\alpha : \alpha \in \Gamma\}$.

Proof. By the remarks preceding the theorem, it is clear that (a) $\Leftrightarrow$ (b) $\Leftrightarrow$ (d). It is easy to verify that a circled, convex, decomposable set is absolutely dominated and that a circled, convex, absolutely dominated set is positively generated. Thus (b) $\Rightarrow$ (c) $\Rightarrow$ (a).

Remark. Similarly we can show that C gives a nearly-open decomposition if and only if the sets of the form $\overline{D(V)}$, where V is a circled, convex $\mathscr{P}$-neighbourhood of 0, constitute a neighbourhood-base at 0.

As suggested by part (b), an ordered convex space with the open decomposition property will be referred to as a *locally decomposable space* and, if it has the nearly-open decomposition property, it will be referred to as a *locally near-decomposable space*. Specializing in *ordered normed spaces* (that is, normed spaces with a vector ordering), we have the following corollary.

(3.2) COROLLARY. *Let* $(E, C, \|.\|)$ *be an ordered normed space and* $\mathscr{P}$ *the vector topology induced by the norm* $\|.\|$. *Then the following statements are equivalent:*

(a) $(E, C, \mathscr{P})$ *is locally decomposable;*

(b) *there exists a norm* $\|.\|_1$ *on E and a positive real number α such that* $\|.\|_1$ *is equivalent to* $\|.\|$ *and C is α-generating in* $(E, \|.\|_1)$ *in the following sense:*

$$\forall x \in E, \ \exists x_1, x_2 \in C \quad such \ that \quad \|x_1\|_1 + \|x_2\|_1 \leqslant \alpha \|x\|_1$$
and
$$x = x_1 - x_2;$$

(c) *there exists a norm* $\|\ \|_2$ *on E which is equivalent to* $\|.\|$ *and has the following property:*

$$\forall \varepsilon > 0, \ \forall x \in E, \ \exists y \in C \quad with \quad \|y\|_2 \leqslant \|x\|_2 + \varepsilon$$
such that
$$-x, x \leqslant y.$$

Proof. (a) $\Rightarrow$ (b): By the corresponding implication in theorem (3.1), there exists a circled, convex, and decomposable $\mathscr{P}$-neighbourhood

V of 0. Let $\|.\|_1$ be the Minkowski functional of V. Then $\|.\|$ and $\|.\|_1$ are equivalent. Also, since V is decomposable, in view of proposition (1.7) (more precisely the proof of (a) $\Rightarrow$ (b) in proposition (1.7)), we conclude that C is $(1+\varepsilon)$-generating in $(E, \|.\|_1)$ for arbitrary $\varepsilon > 0$.

Conversely, if (b) holds, then $\dfrac{1}{\alpha}\Sigma_1 \subseteq \Sigma_1 \cap C - \Sigma_1 \cap C$, where $\Sigma_1 = \{x \in E : \|x\|_1 \leqslant 1\}$. Hence $\Sigma_1 \cap C - \Sigma_1 \cap C$ is a neighbourhood of 0 in $(E, \|\ \|_1)$ and hence in $(E, \mathscr{P})$. This shows that (b) $\Rightarrow$ (a). Therefore (a) $\Leftrightarrow$ (b). Similarly we can show (a) $\Leftrightarrow$ (c).

A cone C in a normed space (or semi-normed space) $(E, \|.\|)$ is said to be α-*generating* if α is a positive constant such that for each x in E there exist x_1, x_2 in C with $\|x_1\| + \|x_2\| \leqslant \alpha \|x\|$ such that $x = x_1 - x_2$. This is the case if and only if $\Sigma \subseteq \alpha D(\Sigma)$, where $\Sigma = \{x \in E : \|x\| \leqslant 1\}$ and $D(\Sigma)$ is the decomposable kernel of Σ in (E, C). C is said to be *strictly generating* if it is α-generating in $(E, \|.\|)$ for some α. It is obvious that a strictly generating cone must be generating. However a generating cone in $(E, \|.\|)$ may not be strict (cf. example (3.7) below).

A cone C in a normed space (or semi-normed space) $(E, \|.\|)$ is said to be a *nearly α-generating* if for each x in E there exist two sequences $\{y_n\}$, $\{z_n\}$ in C such that $\|y_n\| + \|z_n\| \leqslant \alpha\|x\|$ and $\{y_n - z_n\}$ converges to x. Clearly this is the case if and only if $\Sigma \subseteq \alpha\,\overline{D(\Sigma)}$. Similarly to corollary (3.2), the following result can be verified without any difficulty.

(3.3) PROPOSITION. *Let $(E, C, \|.\|)$ be an ordered normed space and $\mathscr{P}$ the vector topology induced by the norm $\|.\|$. Then the following statements are equivalent:*

(a) *C gives a nearly-open decomposition in $(E, \mathscr{P})$;*

(b) *there exists a norm $\|\ \|'$ on E which is equivalent to $\|.\|$ and is such that C is nearly α-generating in $(E, \|.\|')$ for some $\alpha > 0$.*

The following result, due to Nachbin (1965), implies that the open decomposition property, in an ordered metrizable topological vector space, is equivalent to a decomposition property for null sequences.

(3.4) PROPOSITION. *Let $(E, C, \mathscr{P})$ be an ordered topological vector space. Consider the following statements:*

(a) *if $\{x_\alpha\}$ is a net convergent to 0, then, for each α, there exist a_α and b_α in C such that $x_\alpha = a_\alpha - b_\alpha$ and the nets $\{a_\alpha\}$, $\{b_\alpha\}$ converge to 0;*

(b) *statement (a) with 'net' replaced by 'sequence';*

(c) *C gives an open decomposition in $(E, \mathscr{P})$.*

Then (a) *implies* (c), *and* (b) *is equivalent to* (c) *whenever the topology* $\mathscr{P}$
is metrizable.

Proof. Suppose that C does not give an open decomposition in
$(E, \mathscr{P})$. Then there exists a $\mathscr{P}$-neighbourhood U of 0 such that

$$V \not\subseteq C \cap U - C \cap U$$

for every $\mathscr{P}$-neighbourhood V of 0. Let $\mathscr{U}$ be the family of all $\mathscr{P}$-
neighbourhoods of 0. For any $V \in \mathscr{U}$, there exists x_V such that

$$x_V \in V \quad \text{and} \quad x_V \notin C \cap U - C \cap U.$$

It is clear that the net $\{x_V : V \in \mathscr{U}\}$ converges to 0. If $x_V = a_V - b_V$,
where a_V, b_V are in C, then one of a_V, b_V is not in U; hence neither of
the nets $\{a_V : V \in \mathscr{U}\}$, $\{b_V : V \in \mathscr{U}\}$ converges to 0.

If the topology $\mathscr{P}$ is metrizable, then it is clear that (c) implies (b).
This completes the proof.

Given an ordered topological vector space $(E, C, \mathscr{P})$, if $E = C - C$,
then we can construct a new vector topology $\mathscr{P}_\mathrm{D}$ in E with the open
decomposition property. In fact, let $\mathscr{U}$ be a neighbourhood-base at 0
in $(E, \mathscr{P})$ consisting of circled sets. For each $U \in \mathscr{U}$, the set
$U \cap C - U \cap C$ is a circled and absorbing set in E. Then the family
$\{U \cap C - U \cap C : U \in \mathscr{U}\}$ determines uniquely a vector topology,
denoted by $\mathscr{P}_\mathrm{D}$, in E, for which the family is a neighbourhood-base at
0 for $\mathscr{P}_\mathrm{D}$. Clearly $\mathscr{P}_\mathrm{D}$ is the greatest lower bound of all vector topologies
which are finer than $\mathscr{P}$ and have the open decomposition property.
The topology $\mathscr{P}_\mathrm{D}$ will be conveniently referred to as the *vector topology
with the open decomposition property associated with* $\mathscr{P}$. It is clear that $\mathscr{P}$
gives an open decomposition if and only if $\mathscr{P} = \mathscr{P}_\mathrm{D}$. Notice that a
subset B of C is $\mathscr{P}_\mathrm{D}$-bounded if and only if it is $\mathscr{P}$-bounded. If $\mathscr{P}$ is
locally convex then so is $\mathscr{P}_\mathrm{D}$, and $\mathscr{P}_\mathrm{D}$ will also be called the *locally
decomposable topology associated with* $\mathscr{P}$.

(3.5) PROPOSITION. *Let* $(E, C, \mathscr{P})$ *be an ordered topological vector
space such that* $E = C - C$, *and consider the topology* $\mathscr{P}_\mathrm{D}$ *associated with*
$\mathscr{P}$. *Then the following statements hold:*

(a) *a positive subset* B *of* E *is* $\mathscr{P}$-*bounded if and only if it is* $\mathscr{P}_\mathrm{D}$-
bounded;

(b) *if* C *is* $\mathscr{P}$-*closed then a monotone increasing net* $\{x_n\}$ $\mathscr{P}$-*converges
to* x *in* E *if and only if it* $\mathscr{P}_\mathrm{D}$-*converges to* x.

Proof. (a) is easy to verify, and the sufficiency of (b) is trivial since $\mathscr{P} < \mathscr{P}_D$. Conversely, suppose $\{x_n\}$ $\mathscr{P}$-converges to x. Then by proposition (2.1), $x \geqslant x_n$ for each n; and for any circled $\mathscr{P}$-neighbourhood V of 0, there exists n_0 such that $x - x_n \in V$ whenever $n \geqslant n_0$. Consequently, for all $n \geqslant n_0$, it is true that

$$x - x_n \in V \cap C \subseteq V \cap C - V \cap C.$$

This shows that $\{x_n\}$ converges to x in E with respect to $\mathscr{P}_D$.

(3.6) **THEOREM** (Klee). *Let $\mathscr{P}$ be a metrizable vector topology in an ordered vector space $(E,\, C)$ such that $E = C - C$. Then $\mathscr{P}_D$ is also metrizable. If, in addition, C is $\mathscr{P}$-complete, then $(E,\, \mathscr{P}_D)$ is complete. In particular, if C gives an open decomposition in $(E,\, \mathscr{P})$ (and if C is $\mathscr{P}$-complete), then $(E,\, \mathscr{P})$ is complete.*

Proof. Let $\{V_n : n = 1, 2, ...\}$ be a countable base at 0 in $(E,\, \mathscr{P})$ consisting of closed, circled sets such that $V_{n+1} + V_{n+1} \subseteq V_n$ for each n. Then the family of all sets

$$V_n \cap C - V_n \cap C \qquad (n = 1, 2, ...)$$

is a neighbourhood-base at 0 in $(E,\, \mathscr{P}_D)$. Therefore $\mathscr{P}_D$ is metrizable. Next we show that $(E,\, \mathscr{P}_D)$ is complete. In fact, let $\{w_n\}$ be a Cauchy sequence in $(E,\, \mathscr{P}_D)$. Then, for each V_k, there exists $M_k > 0$ such that

$$w_n - w_m \in V_k \cap C - V_k \cap C \quad \text{whenever} \quad m, n \geqslant M_k;$$

therefore, there exists a subsequence $\{z_k\}$ of $\{w_n\}$ such that

$$z_{k+1} - z_k \in V_k \cap C - V_k \cap C \qquad (k = 1, 2, ...).$$

For each k, let $x_k, y_k \in V_k \cap C$ be such that

$$z_{k+1} - z_k = x_k - y_k.$$

Then

$$z_{n+1} - z_1 = \sum_{k=1}^{n} (z_{k+1} - z_k) = \sum_{k=1}^{n} x_k - \sum_{k=1}^{n} y_k.$$

To show the convergence of the $\mathscr{P}_D$-Cauchy sequence $\{w_n\}$, it is sufficient to show the convergence of the subsequence $\{z_n\}$, and in turn it is sufficient to show the convergence of the formal series $\sum x_k$ and $\sum y_k$ in $(E,\, \mathscr{P}_D)$. Let $u_n = \sum_{k=1}^{n} x_k$, for each positive integer n. Then each $u_n \in C$ and

$$u_{n+q} - u_n = x_{n+1} + ... + x_{n+q} \in V_{n+1} \cap C + ... + V_{n+q} \cap C \subseteq V_n \cap C;$$

this shows that $\{u_n\}$ is a $\mathcal{P}$-Cauchy sequence in C. Since C is $\mathcal{P}$-complete, $\{u_n\}$ converges to an element, say u, in C. Further, since V_n and C are $\mathcal{P}$-closed, passing to the limit as $q \to \infty$ in the last displayed formula we see that $u - u_n \in V_n \cap C \subseteq V_n \cap C - V_n \cap C$. This shows that the series $\sum\limits_{k=1}^{\infty} x_k$ converges in $(E, \mathcal{P}_{\mathrm{D}})$ to u. Similarly, we can show that $\sum\limits_{k=1}^{\infty} y_k$ converges. The proof of the theorem is complete.

We have noted earlier that an ordered topological vector space $(E, C, \mathcal{P})$ has the open decomposition property if and only if $\mathcal{P} = \mathcal{P}_{\mathrm{D}}$, and that this property certainly implies C is a generating cone. However, the converse is incorrect as the following example shows.

(3.7) EXAMPLE. Consider the Banach space $C[0, 1]$ of all real-valued continuous functions defined on $[0, 1]$. Let C be the cone in $C[0, 1]$ consisting of all non-negative and convex functions, and let $E = C - C$. By the Stone–Weierstrass theorem, E is dense in the Banach space $C[0, 1]$. Also, since any function in C must be differentiable on $[0, 1]$ except at, at most, a countable number of points, any function f in $C[0, 1]$ which is not differentiable at an uncountable subset of $[0, 1]$ is not in E. Therefore E is a proper dense subspace of $C[0, 1]$, and E is a non-complete normed space in its own right. Further, C is a generating and norm-complete cone in E. By the preceding theorem of Klee, the cone C does not give an open decomposition in $(E, \|.\|)$.

The following result implies however that in a *Banach* space, a generating and complete cone always gives an open decomposition.

(3.8) THEOREM. *Let $(E, C, \mathcal{P})$ be a metrizable and ordered topological vector space, and suppose that C is $\mathcal{P}$-complete. Then the following statements* (a) *and* (b) *are equivalent:*

(a) $(E, C, \mathcal{P})$ *has the open decomposition property;*

(b) $(E, C, \mathcal{P})$ *has the nearly-open decomposition property.*

Further, if E is of the second category, then each of the statements is equivalent to

(c) C *is generating in E,*

and, in this case, E must be $\mathcal{P}$-complete.

Proof. It is trivial that (a) $\Rightarrow$ (b). Conversely, suppose (b) holds. Let $F = C - C$ and let $\mathcal{P}_{\mathrm{D}}$ be the vector topology in F with the open decomposition property associated with $\mathcal{P}$. By Klee's theorem (theorem (3.6)), $(F, \mathcal{P}_{\mathrm{D}})$ is metrizable and complete. Let i be the identity map from $(F, \mathcal{P}_{\mathrm{D}})$ into $(E, \mathcal{P})$. Then i is continuous. Further, take a $\mathscr{P}_{\mathrm{D}}$-neighbourhood of 0 in F of the form $V \cap C - V \cap C$, where V is a $\mathscr{P}$-neighbourhood of 0 in E. Then, by (b), $\overline{i(V \cap C - V \cap C)}$ is a neighbourhood of 0 in $(E, \mathscr{P})$. This shows that the continuous map i is 'nearly-open'. By the Banach open-mapping theorem (cf. Schaefer (1966, p. 76)), i is open and hence i is a homeomorphism from $(F, \mathscr{P}_{\mathrm{D}})$ onto $(E, \mathscr{P})$. In particular, $\mathscr{P} = \mathscr{P}_{\mathrm{D}}$ and C gives an open decomposition in $(E, \mathscr{P})$. Therefore (b) $\Rightarrow$ (a) and statements (a) and (b) are equivalent.

It is trivial that (a) $\Rightarrow$ (c). Conversely, we show that (c) $\Rightarrow$ (a) for the case when $(E, \mathscr{P})$ is of the second category. By (c), $E = C - C = F$, so i is a continuous linear mapping from the complete metrizable space $(E, \mathscr{P}_{\mathrm{D}})$ onto the space $(E, \mathscr{P})$ of the second category. By the open-mapping theorem (cf. Schaefer (1966, p. 76)), i is in fact a homeomorphism; hence $\mathscr{P} = \mathscr{P}_{\mathrm{D}}$ and C gives an open decomposition in $(E, \mathscr{P})$. Further, since E is $\mathscr{P}_{\mathrm{D}}$-complete, so is $\mathscr{P}$-complete. This completes the proof of the theorem.

Let us consider the case when $\mathscr{P}$ is a locally convex topology in an ordered vector space (E, C) such that $E = C - C$. Let $\mathscr{U}$ be a neighbourhood-base at 0 consisting of circled *convex* sets. Since, for all $V \in \mathscr{U}$,

$$\tfrac{1}{2}(V \cap C - V \cap C) \subseteq D(V) \subseteq V \cap C - V \cap C,$$

the family $D(\mathscr{U}) = \{D(V) : V \in \mathscr{U}\}$ is a neighbourhood-base at 0 in E with respect to the associated locally decomposable topology $\mathscr{P}_{\mathrm{D}}$. In particular, if $\mathscr{P}$ is the vector topology induced by a norm $\|.\|$ and if $\Sigma = \{x \in E : \|x\| \leqslant 1\}$, then $\{\varepsilon D(\Sigma) : \varepsilon > 0\} = \{D(\varepsilon\Sigma) : \varepsilon > 0\}$ is a neighbourhood-base at 0 in $(E, \mathscr{P}_{\mathrm{D}})$. Thus $\mathscr{P}_{\mathrm{D}}$ is precisely the vector topology induced by the Minkowski functional of $D(\Sigma)$. Specializing in this normable case, we have the following numerical version of the preceding theorem.

(3.9) **THEOREM.** *Let $(E, C, \|.\|)$ be an ordered normed space and suppose that C is $\|.\|$-complete. Let α be a positive constant. Then the following statements are equivalent:*

(a) *C is $(\alpha + \varepsilon)$-generating for each $\varepsilon > 0$;*

(b) *C is nearly α-generating.*

Further, if (a) *or* (b) *holds, then $(E, \|.\|)$ is complete.*

Proof. Let Σ denote the closed unit ball in $(E, \|.\|)$, and let $D(\Sigma)$ denote the decomposable kernel of Σ. Define a new norm $\|.\|'$ in $F = C - C$ to be the Minkowski functional of $D(\Sigma)$. Then $(F, \|.\|')$ is a

Banach space, by Klee's theorem. Also, the identity map i from $(E, \|\,.\,\|')$ into $(E, \|\,.\,\|)$ is continuous. By (b), $\Sigma \subseteq \alpha\, \overline{D(\Sigma)}$, so the closure of the image of $D(\Sigma)$ under i is a neighbourhood of 0 in $(E, \|\,.\,\|)$. By a well-known result (*cf.* Schaefer (1966, p. 76)),

$$i(D(\beta\Sigma)) \supseteq \overline{i(D(\gamma\Sigma))},$$

whenever $\beta > \gamma > 0$. In particular, in the space $(E, \|\,.\,\|)$,

$$(\alpha+\varepsilon)D(\Sigma) \supseteq \alpha\, \overline{D(\Sigma)} \supseteq \Sigma \qquad (\varepsilon > 0),$$

which implies that C is $(\alpha+\varepsilon)$-generating for each $\varepsilon > 0$. This shows that (b) $\Rightarrow$ (a). Conversely, if (a) holds then

$$\Sigma \subseteq \alpha\, D(\Sigma) + \varepsilon\, D(\Sigma) \subseteq \alpha\, D(\Sigma) + \varepsilon\Sigma \qquad (\varepsilon > 0);$$

hence $\Sigma \subseteq \alpha\, \overline{D(\Sigma)}$, proving (b).

Returning to the case when $(E, C, \mathscr{P})$ is an ordered convex space, the following theorem gives a dual characterization of the $\mathscr{P}_{\mathrm{D}}$-topology. We first recall (cf. theorem (2.11)) that if V is a circled convex $\mathscr{P}$-neighbourhood of 0 then the polar V^{π}, taken in E^*, is the same as the polar V^0 taken in E', and $(D(V))^{\pi} = F(V^0)$, where $F(V^0)$ denotes the order-convex hull of V^0 in E^*. Since $D(\mathscr{U}) = \{D(V)\colon V \in \mathscr{U}\}$ is a neighbourhood-base at 0 in $(E, \mathscr{P}_{\mathrm{D}})$, it follows that $\mathscr{P}_{\mathrm{D}}$ is the topology of uniform convergence on all sets of the form $F(V^0)$, i.e. on the order-convex hulls in E^* of $\mathscr{P}$-equicontinuous subsets of E'. Thus we arrive at the following theorem.

(3.10) THEOREM. *Let $(E, C, \mathscr{P})$ be an ordered convex space such that $E = C - C$. Then the locally decomposable topology $\mathscr{P}_{\mathrm{D}}$ associated with $\mathscr{P}$ is the topology of uniform convergence on the order-convex hulls in the algebraic dual (E^*, C^*) of all $\mathscr{P}$-equicontinuous subsets of E'. Consequently $\mathscr{P}$ is locally decomposable (that is $\mathscr{P} = \mathscr{P}_{\mathrm{D}}$) if and only if the order-convex hull in E^* of each $\mathscr{P}$-equicontinuous subset is $\mathscr{P}$-equicontinuous.*

Similarly, we have the following dual characterization (Duhoux 1972b) for $(E, C, \mathscr{P})$ to have the nearly-open decomposition property.

(3.11) THEOREM. *Let $(E, C, \mathscr{P})$ be an ordered convex space. Then C gives a nearly-open decomposition in $(E, \mathscr{P})$ if and only if the order-convex hull in E' of each $\mathscr{P}$-equicontinuous subset of E' is $\mathscr{P}$-equicontinuous.*

Proof. (a) *Necessity.* Let A be a $\mathscr{P}$-equicontinuous subset of E'. Then there exists a circled convex $\mathscr{P}$-neighbourhood V of 0 such that $A \subseteq V^0 = V^\pi$. By part (a) of the duality theorem (2.11), we have

$$\overline{(D(V))^0} = (D(V))^0 = (D(V))^\pi \cap E' = F(V^0) \cap E' \supseteq F(A) \cap E'$$

where $F(A)$ denotes the order-convex hull in E^* of A (so $F(A) \cap E'$ is the order-convex hull in E' of A). Since C gives a nearly-open decomposition, $\overline{D(V)}$ is a $\mathscr{P}$-neighbourhood of 0, it follows that the order-convex hull in E' of A is $\mathscr{P}$-equicontinuous.

(b) *Sufficiency.* Let U be a circled convex $\mathscr{P}$-neighbourhood of 0. Then U^0 is $\mathscr{P}$-equicontinuous and hence $F(U^0) \cap E'$ is $\mathscr{P}$-equicontinuous by assumption; i.e., $\overline{(D(U))^0}$ is $\mathscr{P}$-equicontinuous by theorem (2.11). Hence $\overline{D(U)}$ is a $\mathscr{P}$-neighbourhood of 0. This shows that C gives a nearly-open decomposition.

Theorems (3.10) and (3.11) have many important applications; we mention a few below.

(3.12) COROLLARY. *Let $(E, C, \mathscr{P})$ be an ordered convex space and $\mathscr{P}_{\mathrm{D}}$ the locally decomposable topology in E associated with $\mathscr{P}$. Then the topological dual $(E, \mathscr{P}_{\mathrm{D}})'$ of $(E, \mathscr{P}_{\mathrm{D}})$ is equal to the order-convex hull in E^* of $E' = (E, \mathscr{P})'$. In particular, if $\mathscr{P}$ is locally decomposable then E' is order-convex in E^*.*

Proof. Apply theorem (3.10).

The following result should be compared with theorem (3.8).

(3.13) COROLLARY. *For an ordered convex space $(E, C, \mathscr{P})$, the following statements are equivalent:*

(a) *C gives an open decomposition in $(E, \mathscr{P})$;*

(b) *C gives a nearly-open decomposition in $(E, \mathscr{P})$ and E' is order-convex in E^*.*

Proof. In view of the preceding corollary it is clear that (a) $\Rightarrow$ (b). Conversely, suppose (b) holds. Then, by theorem (3.11), the order-convex hull in E' of each $\mathscr{P}$-equicontinuous subset A of E' is equicontinuous. Since E' itself is order-convex in E^*, the order-convex hull of A in E' is the same as the order-convex hull in E^*. Thus the order-convex hull in E^* of each $\mathscr{P}$-equicontinuous subset of E' is $\mathscr{P}$-equicontinuous. By theorem (3.10), $\mathscr{P}$ must be locally decomposable; this shows that (b) $\Rightarrow$ (a).

(3.14) COROLLARY. *Let $(E, C, \mathscr{P})$ be an ordered convex space with the topological dual E' and suppose that $E = C - C$. Let $\tau(E, E') \equiv \tau$ be the Mackey topology in E. Then $\tau(E, E')$ is locally decomposable if and only if E' is order-convex in E^*. In particular, if $(E, C, \mathscr{P})$ is locally decomposable then so is $(E, C, \tau(E, E'))$.*

Proof. We recall that $\tau(E, E')$ is the strongest locally convex topology in E with the dual E'. Thus, if τ is locally decomposable then E' is order-convex in E^* by corollary (3.12). Conversely, suppose E' is order-convex in E^*, and let τ_{D} be the locally decomposable topology associated with τ. By corollary (3.12) again, the dual of (E, τ_{D}) coincides with E'. Since $\tau \leqslant \tau_{\mathrm{D}}$ it follows from the definition of the Mackey topology that $\tau = \tau_{\mathrm{D}}$; hence τ is locally decomposable. The last assertion of this corollary follows from the first and from corollary (3.12).

We now give an example of an ordered convex space $(E, C, \mathscr{P})$ for which E' is order-convex in E^* but $\mathscr{P}$ is *not* locally decomposable (though the Mackey topology $\tau(E, E')$ is locally decomposable). This, in particular, shows that the open decomposition property is *not* invariant with respect to all topologies for a dual pair $\langle E, E' \rangle$.

(3.15) EXAMPLE. Let $L^1 = L^1[0, 1]$ be the ordered Banach space of all Lebesgue integrable real-valued functions on $[0, 1]$ with the usual L^1-norm $\|.\|$ and the usual ordering (we identify, as usual, the functions which are equal almost everywhere). Then the topological dual of L^1 is $L^\infty = L^\infty[0, 1]$. It is well known that the norm topology in L^1 is precisely the Mackey topology $\tau(L^1, L^\infty)$, and is strictly finer than the weak topology $\sigma(L^1, L^\infty)$. Let V be the polar in L^1 of the constant function 1 (as an element of L^∞). Then V is a $\sigma(L^1, L^\infty)$-neighbourhood of 0. Let Σ be the closed unit ball in L^1. Then it is easy to verify that

$$V \cap C - V \cap C \subseteq 2\Sigma,$$

where C denotes the positive cone in L^1. Since Σ is *not* a $\sigma(L^1, L^\infty)$-neighbourhood of 0, $V \cap C - V \cap C$ must not be a neighbourhood of 0. Therefore C does not have the open decomposition property in $(L^1, \sigma(L^1, L^\infty))$.

Finally, we consider some permanence property of locally decomposable spaces. Let $\{(E_\alpha, C_\alpha, \mathscr{P}_\alpha) : \alpha \in \Gamma\}$ be a family of ordered convex spaces. Let $E = \prod_{\alpha \in \Gamma} E_\alpha$ denote the Cartesian product of E_α, ordered by the product cone $C = \prod_{\alpha \in \Gamma} C_\alpha$. Let $F = \bigoplus_{\alpha \in \Gamma} E_\alpha$ be the algebraic direct sum of the E_α, ordered by relative cone $C \cap F$. For each $\alpha \in \Gamma$, j_α denotes the natural canonical embedding map from E_α into F. The most important topology for F is perhaps the so-called *locally convex direct-sum topology* $\bigoplus_{\alpha \in \Gamma} \mathscr{P}_\alpha$, which is the finest locally convex topology for which j_α is continuous. A neighbourhood-base at 0 for this topology is provided by the family of all sets of the form $V = \mathrm{co}(\bigcup_{\alpha \in \Gamma} V_\alpha)$, where each V_α is a neighbourhood of 0 in $(E_\alpha, \mathscr{P}_\alpha)$. Here, as usual, we write V_α instead of $j_\alpha(V_\alpha)$; thus each E_α is considered as a subspace of F.

(3.16) **THEOREM.** *Let $\mathscr{P}_{\alpha,\mathrm{D}}$ be the locally decomposable topology associated with $\mathscr{P}_\alpha$, and let $(\bigoplus_{\alpha \in \Gamma} \mathscr{P}_\alpha)_\mathrm{D}$ be the locally decomposable topology associated with $\bigoplus_{\alpha \in \Gamma} \mathscr{P}_\alpha$. Then*

$$\left(\bigoplus_{\alpha \in \Gamma} \mathscr{P}_\alpha \right)_\mathrm{D} = \bigoplus_{\alpha \in \Gamma} \mathscr{P}_{\alpha,\mathrm{D}}. \tag{3.1}$$

Consequently the following assertions hold:

    (a) *if each $\mathscr{P}_\alpha$ is locally decomposable then so is $\bigoplus_{\alpha \in \Gamma} \mathscr{P}_\alpha$;*

    (b) *the order-convex hull in F^* of the product $\prod_{\alpha \in \Gamma} (E_\alpha, \mathscr{P}_\alpha)'$ is precisely the product of order-convex hulls in E_α^* of $(E_\alpha, \mathscr{P}_\alpha)'$.*

Proof. For each α in Γ, let V_α be a circled convex neighbourhood of 0 in $(E_\alpha, \mathscr{P}_\alpha)$. Let

$$A = \mathrm{co}\left(\bigcup_{\alpha \in \Gamma} V_\alpha \right) \quad \text{and} \quad B = \mathrm{co}\left(\bigcup_{\alpha \in \Gamma} D(V_\alpha) \right).$$

To prove the formula (3.1), it is sufficient to show that $D(A) = B$, because $D(A)$ and B are 'typical' neighbourhoods of 0 in F with respect to $(\bigoplus_{\alpha \in \Gamma} \mathscr{P}_\alpha)_\mathrm{D}$ and $\bigoplus_{\alpha \in \Gamma} \mathscr{P}_{\alpha,\mathrm{D}}$, respectively.

It is easy to see that B is a decomposable set; hence to show $D(A) = B$ it suffices to show that $A \cap C = B \cap C$. Since $A \supseteq B$, we have only to show that $A \cap C \subseteq B \cap C$. Let $x \in A \cap C$ and suppose that $x = \sum_{i=1}^{n} \lambda_i x_i$, where $\sum_{i=1}^{n} \lambda_i = 1$, each $\lambda_i > 0$, $x_i \in V_{\alpha_i}$, and $\alpha_i \in \Gamma$. By the definition of the product cone C, it follows from $x \in C$ that each $x_i \in C_{\alpha_i}$. Hence $x_i \in V_{\alpha_i} \cap C_{\alpha_i} \subseteq D(V_{\alpha_i})$, and $x \in B$. This shows that $x \in B \cap C$ and hence that $A \cap C \subseteq B \cap C$. Therefore formula (3.1) is proved, and assertion (a) follows immediately. The assertion (b) also follows from formula (3.1) and corollary (3.12) together with the following well-known result:

$$\left(F, \bigoplus_{\alpha\in\Gamma} \mathscr{P}_\alpha\right)' = \prod_{\alpha\in\Gamma} (E_\alpha, \mathscr{P}_\alpha)' \quad\text{and}\quad \left(F, \bigoplus_{\alpha\in\Gamma} \mathscr{P}_{\alpha,\mathrm{D}}\right)' = \prod_{\alpha\in\Gamma} (E_\alpha, \mathscr{P}_{\alpha,\mathrm{D}})'.$$

The assertion (a) in the preceding theorem can be further generalized in the following form. (A map from an ordered vector space into another space is *positive* if it sends positive elements into positive elements.)

(3.17) THEOREM. *Let (E, C) be an ordered vector space, $\{(E_\alpha, C_\alpha, \mathscr{P}_\alpha): \alpha \in \Gamma\}$ a family of locally decomposable spaces, and t_α a positive linear map from E_α into E. Suppose that E is the linear hull of $\bigcup_{\alpha\in\Gamma} t_\alpha(E_\alpha)$. Then the inductive topology $\mathscr{P}$ on E with respect to $\{(E_\alpha, C_\alpha, \mathscr{P}_\alpha): \alpha \in \Gamma\}$ and $\{t_\alpha: \alpha \in \Gamma\}$ is locally decomposable.*

Proof. Recall that a neighbourhood-base at 0 for $\mathscr{P}$ is provided by the family of all sets of the form $V = \mathrm{co}\left(\bigcup_{\alpha\in\Gamma} t_\alpha(V_\alpha)\right)$, where each V_α is a neighbourhood of 0 in $(E_\alpha, \mathscr{P}_\alpha)$. Now, since $\mathscr{P}_\alpha$ is locally decomposable, we can take V_α to be circled, convex, and decomposable. Since t_α is positive, $t_\alpha(V_\alpha)$ must also be decomposable. Consequently V, as the convex hull of decomposable sets, must be decomposable. This implies that $\mathscr{P}$ is locally decomposable.

(3.18) COROLLARY. *Let $(E, C, \mathscr{P})$ be a locally decomposable space, J an order-convex subspace of E, and let q be the quotient map from E onto E/J. Then the quotient topology is locally decomposable in $(E/J, C/J)$.*

The following result may be regarded as a dual to theorem (3.16).

(3.19) THEOREM. *Let $\{(E_\alpha, C_\alpha, \mathscr{P}_\alpha): \alpha \in \Gamma\}$ be a family of ordered convex spaces, and let E be the product space ordered by the product cone C. Let $\prod_{\alpha\in\Gamma} \mathscr{P}_\alpha$ be the product topology and $\left(\prod_{\alpha\in\Gamma} \mathscr{P}_\alpha\right)_\mathrm{D}$ the locally decomposable topology associated with $\prod_{\alpha\in\Gamma} \mathscr{P}_\alpha$. Then*

$$\left(\prod_{\alpha\in\Gamma} \mathscr{P}_\alpha\right)_\mathrm{D} = \prod_{\alpha\in\Gamma} \mathscr{P}_{\alpha,\mathrm{D}}. \tag{3.2}$$

Consequently the following assertions hold:

(a) *if each $\mathscr{P}_\alpha$ is locally decomposable then so is $\prod \mathscr{P}_\alpha$;*

(b) *the order-convex hull in E^* of the direct sum $\bigoplus\limits_{\alpha \in \Gamma} (E_\alpha, \mathscr{P}_\alpha)'$ is precisely the direct sum of order-convex hulls in E_α^* of $(E_\alpha, \mathscr{P}_\alpha)'$.*

Proof. For each α in Γ, let V_α be a circled convex neighbourhood of 0 in $(E_\alpha, \mathscr{P}_\alpha)$, and suppose all V_α except for a finite number of α in Γ are equal to E_α. Then the formula (3.2) follows from an easily verified fact:

$$\left(\prod_{\alpha \in \Gamma} V_\alpha\right) \cap C - \left(\prod_{\alpha \in \Gamma} V_\alpha\right) \cap C = \prod_{\alpha \in \Gamma} (V_\alpha \cap C - V_\alpha \cap C),$$

because the set on the left-hand side is a 'typical' neighbourhood of 0 with respect to $\left(\prod\limits_{\alpha \in \Gamma} \mathscr{P}_\alpha\right)_{\mathrm{D}}$-topology and the set on the right-hand side is the same with respect to $\left(\prod\limits_{\alpha \in \Gamma} \mathscr{P}_{\alpha,\mathrm{D}}\right)$-topology.

Assertion (a) follows immediately from formula (3.2); and assertion (b) follows also from formula (3.2) and corollary (3.12) together with the following well-known result:

$$\left(E, \prod_{\alpha \in \Gamma} \mathscr{P}_\alpha\right)' = \bigoplus_{\alpha \in \Gamma} (E, \mathscr{P}_\alpha)'$$

and

$$\left(E, \prod_{\alpha \in \Gamma} \mathscr{P}_{\alpha,\mathrm{D}}\right)' = \bigoplus_{\alpha \in \Gamma} (E_\alpha, \mathscr{P}_{\alpha,\mathrm{D}})'.$$

4

$\mathscr{B}$-CONES AND LOCAL $\mathscr{B}$-CONES

CONSIDER a non-empty set X and let $\mathscr{C}$ be a family of subsets of X. A subfamily $\mathscr{C}_0$ of $\mathscr{C}$ is called a *fundamental system* (or a *fundamental subfamily*) for $\mathscr{C}$ if each member of $\mathscr{C}$ is contained in some member of $\mathscr{C}_0$; in other words, if $\mathscr{C}_0$ is cofinal in $\mathscr{C}$ under the set inclusion relation.

Let $(E, C, \mathscr{P})$ be an ordered topological vector space, and $\mathscr{B}$ the family of all $\mathscr{P}$-bounded subsets of E. Let $\mathscr{C} \subseteq \mathscr{B}$. The positive cone C is called an $\mathscr{C}$-*cone* if the family

$$\overline{\{A \cap C - A \cap C : A \in \mathscr{C}\}}$$

is a fundamental system for $\mathscr{C}$, and C is called a *strict $\mathscr{C}$-cone* if the family $\{A \cap C - A \cap C : A \in \mathscr{C}\}$ is a fundamental system for $\mathscr{C}$. In particular, C is an $\mathscr{B}$-*cone* if each $\mathscr{P}$-bounded subset of E is contained in a set of the form $\overline{A \cap C - A \cap C}$ where A is a $\mathscr{P}$-bounded set in E, and C is a *strict $\mathscr{B}$-cone* if each $\mathscr{P}$-bounded subset of E is contained in a set of the form $A \cap C - A \cap C$. The former is the case if and only if the family of all closures of all positively generated $\mathscr{P}$-bounded subsets is fundamental in $\mathscr{B}$, and the latter is the case if and only if the family of all positively generated $\mathscr{P}$-bounded subsets is fundamental in $\mathscr{B}$. In view of the Mackey–Arens theorem, if $\mathscr{P}_1$, $\mathscr{P}_2$ are two locally convex topologies in E with the same topological dual, then a cone C in E is an $\mathscr{B}$-cone (or strict $\mathscr{B}$-cone) with respect to $\mathscr{P}_1$ if and only if it is so with respect to $\mathscr{P}_2$.

Let V be a neighbourhood of 0 in an ordered topological vector space $(E, C, \mathscr{P})$. A subset B of E is said to be *locally bounded with respect to V* if B is absorbed by V, i.e. if there exists $\lambda > 0$ such that $B \subseteq \lambda V$. The positive cone C is called a *locally strict $\mathscr{B}$-cone* if, for any $\mathscr{P}$-bounded subset B of E and any $\mathscr{P}$-neighbourhood V of 0, there exist two positive subsets B_1, B_2 which are locally bounded with respect to V such that $B \subseteq B_1 - B_2$. C is called a *local $\mathscr{B}$-cone* if, for any $\mathscr{P}$-bounded subset B of E and any $\mathscr{P}$-neighbourhood V of 0, there exist two positive subsets B_1, B_2 which are locally bounded with respect to V such that $B \subseteq \overline{B_1 - B_2}$. Actually in the decomposition we can always take $B_1 = B_2$ (replace $B_1 \cup B_2$ for B_1, B_2 if necessary). Moreover, since each $\mathscr{P}$-bounded set is locally bounded with respect to each neighbourhood of 0, it is obvious that each $\mathscr{B}$-cone must be a local $\mathscr{B}$-cone and each strict $\mathscr{B}$-cone must be a locally strict $\mathscr{B}$-cone. In the case when $\mathscr{P}$ is normable then the converse also holds.

(4.1) PROPOSITION. *Let $(E, C, \mathscr{P})$ be an ordered topological vector space. If C gives an open decomposition then C must be a locally strict $\mathscr{B}$-cone. Conversely, in the case when $(E, \mathscr{P})$ is a bornological locally convex space, any locally strict $\mathscr{B}$-cone (and, in particular, any strict $\mathscr{B}$-cone) in E gives an open decomposition.*

Proof. Suppose $(E, C, \mathscr{P})$ has the open decomposition property. Let B be a $\mathscr{P}$-bounded subset of E, and V a $\mathscr{P}$-neighbourhood of 0. Then $V \cap C - V \cap C$ is also a $\mathscr{P}$-neighbourhood of 0 and thus absorbs B; i.e. there exists $\lambda > 0$ such that $B \subseteq \lambda(V \cap C - V \cap C)$. Let $A = \lambda(V \cap C)$. Then A is a positive subset of E, absorbed by V and $B \subseteq A - A$, showing that C is a locally strict $\mathscr{B}$-cone. Conversely, suppose C is a locally strict $\mathscr{B}$-cone in $(E, \mathscr{P})$ and that $(E, \mathscr{P})$ is a bornological (locally convex) space. Let V be a circled convex $\mathscr{P}$-neighbourhood of 0. We have to show that $V \cap C - V \cap C$ is a $\mathscr{P}$-neighbourhood of 0. To this end, let us take a $\mathscr{P}$-bounded set B in E. Then there exist *positive* subsets B_1, B_2 of E such that $B \subseteq B_1 - B_2$ and B_1, B_2 are absorbed by V. Hence B is absorbed by $V \cap C - V \cap C$. It is now clear that $V \cap C - V \cap C$ is a circled convex set in E and absorbs all $\mathscr{P}$-bounded subsets. Since $\mathscr{P}$ is bornological, it follows that $V \cap C - V \cap C$ is a $\mathscr{P}$-neighbourhood of 0.

Remark. By example (3.15), the bornological assumption in the preceding proposition cannot be dropped.

Similarly, we can show the following proposition to be true.

(4.2) PROPOSITION. *Let $(E, C, \mathscr{P})$ be an ordered topological vector space. If C gives a nearly-open decomposition then C must be a local $\mathscr{B}$-cone. Conversely, in the case when $(E, \mathscr{P})$ is an infrabarrelled locally convex space, any local $\mathscr{B}$-cone (and, in particular, any $\mathscr{B}$-cone) in E gives a nearly-open decomposition.*

Proof. Similar to that given in the proof of proposition (4.1).

(4.3) PROPOSITION. *Let $(E, C, \mathscr{P})$ be an ordered topological vector space and suppose that $E = C - C$. Let $\mathscr{P}_{\mathrm{D}}$ be the topology in E associated* with $\mathscr{P}$ with the open decomposition property. Then the following statements are equivalent:

(a) any subset B of E is $\mathscr{P}_D$-bounded if (and only if) it is $\mathscr{P}$-bounded;

(b) C is a locally strict $\mathscr{B}$-cone in $(E, \mathscr{P})$.

Proof. (a) $\Rightarrow$ (b): Let B be a $\mathscr{P}$-bounded subset of E. Then B is $\mathscr{P}_D$-bounded by (a). Further, by the first part of proposition (4.1), C is a locally strict $\mathscr{B}$-cone in $(E, \mathscr{P}_D)$. Hence, for any $\mathscr{P}$-neighbourhood V of 0, there exist positive subsets B_1, B_2 of E such that $B \subseteq B_1 - B_2$ and each of B_1, B_2 is absorbed by the $\mathscr{P}_D$-neighbourhood $V \cap C - V \cap C$ of 0, and in particular absorbed by V. Therefore (b) holds.

(b) $\Rightarrow$ (a): Let B be a $\mathscr{P}$-bounded subset of E and let

$$V \cap C - V \cap C = U$$

be a $\mathscr{P}_D$-neighbourhood of 0, where V is a $\mathscr{P}$-neighbourhood of 0. We have to show that B is absorbed by U. By (b), there exist two positive sets B_1, B_2 such that $B \subseteq B_1 - B_2$ and each of B_1, B_2 is absorbed by V and hence by $V \cap C$ (since B_1, B_2 are positive) and *a fortiori* by U. This implies that B is $\mathscr{P}_D$-bounded.

(4.4) PROPOSITION. *Let $(E, C, \mathscr{P})$ be an ordered convex space and $\mathscr{B}$ the family of all $\mathscr{P}$-bounded sets. Then the following statements are equivalent:*

(a) *C is a strict $\mathscr{B}$-cone in $(E, \mathscr{P})$;*

(b) *for each $\mathscr{P}$-bounded subset B of E there exists a circled convex, decomposable, and $\mathscr{P}$-bounded subset A of E such that $B \subseteq A$;*

(c) *for each $\mathscr{P}$-bounded subset B of E there exists a circled convex, absolute-dominated, and $\mathscr{P}$-bounded subset A of E such that $B \subseteq A$.*

(a), (b), (c) *are also equivalent if $B \subseteq A$ is replaced by $B \subseteq \bar{A}$ in (b) and (c) and in (a) C is a $\mathscr{B}$-cone.*

Proof. Recall that if A is a circled convex subset of E then the decomposable kernel $D(A)$ is absolutely dominated and

$$\tfrac{1}{2}(A \cap C - A \cap C) \subseteq D(A) \subseteq A \cap C - A \cap C.$$

Hence (a) $\Leftrightarrow$ (b) and (b) $\Rightarrow$ (c). On the other hand, if A is an absolutely dominated, circled convex set then A must be positively generated; hence (c) $\Rightarrow$ (a).

(4.5) PROPOSITION. *Let* $(E, C, \mathscr{P})$ *be an ordered convex space. Then the following statements are equivalent:*

(a) *C is a locally strict $\mathscr{B}$-cone in* $(E, \mathscr{P})$;

(b) *for each $\mathscr{P}$-bounded subset B of E and each $\mathscr{P}$-neighbourhood V of 0 there exists a circled convex and decomposable subset A of E such that A is absorbed by V and $B \subseteq A$;*

(c) *for each $\mathscr{P}$-bounded subset B of E and each $\mathscr{P}$-neighbourhood V of 0 there exist a circled convex and absolute-dominated subset A of E such that A is absorbed by V and $B \subseteq A$.*

(a), (b), (c) *are also equivalent if $B \subseteq A$ is replaced by $B \subseteq \bar{A}$ in* (b) *and* (c) *and in* (a) *C is a local $\mathscr{B}$-cone.*

Proof. Similar to that given for proposition (4.4).

(4.6) PROPOSITION. *For any ordered normed space* $(E, C, \|.\|)$, *the following statements are equivalent:*

(a) *C gives an open decomposition in* $(E, \|.\|)$;

(b) *C is a locally strict $\mathscr{B}$-cone in* $(E, \|.\|)$;

(c) *C is a strict $\mathscr{B}$-cone in* $(E, \|.\|)$;

(d) *there exists a positive real number α such that C is α-generating in* $(E, \|.\|)$.

The following statements are also equivalent:

(a)' *C gives a nearly-open decomposition in* $(E, \|.\|)$;

(b)' *C is a local $\mathscr{B}$-cone in* $(E, \|.\|)$;

(c)' *C is a $\mathscr{B}$-cone in* $(E, \|.\|)$;

(d)' *there exists a positive real number α such that C is nearly α-generating in* $(E, \|.\|)$.

Proof. The equivalence (a) $\Leftrightarrow$ (d) was established in corollary (3.2). That (a) $\Leftrightarrow$ (b) was proved in proposition (4.1). Similarly, it is easy to verify that (b) $\Leftrightarrow$ (c). The equivalence of (a)', (b)', (c)', (d)' follows similarly.

If $(E, \|.\|)$ is of the second category (e.g. if E is a Banach space) we can improve the implications (d) $\Rightarrow$ (a) and (d)' $\Rightarrow$ (a)' as follows.

(4.7) PROPOSITION. *Let* $(E, C, \|.\|)$ *be an ordered normed space which is of the second category. If for each x in E there exist two norm-bounded sequences $\{y_n\}, \{z_n\}$ in C such that $\{y_n - z_n\}$ converges to x in* $(E, \|.\|)$ *(in particular, if $E = C - C$), then C is a $\mathscr{B}$-cone in* $(E, \|.\|)$ *and hence is nearly α-generating for some $\alpha > 0$. If, in addition, C is $\|.\|$-complete, then C is $(\alpha + \varepsilon)$-generating for all $\varepsilon > 0$.*

Proof. Let Σ be the closed unit ball in $(E, \|.\|)$, and let $D(\Sigma)$ be the decomposable kernel of Σ. Then the stated decomposition assumption of the proposition implies that $E = \bigcup_{n=1}^{\infty} n\overline{D(\Sigma)}$. Since E is of the second category, there must exist some positive integer N such that $N\overline{D(\Sigma)}$ has a non-empty interior. Since $N\,\overline{D(\Sigma)}$ is circled convex, its interior must contain the origin. In other words, $N\,\overline{D(\Sigma)}$ is a neighbourhood of 0, and so is $\overline{D(\Sigma)}$. This shows that C gives a nearly-open decomposition in $(E, \|.\|)$ and hence, C is nearly α-generating for some $\alpha > 0$. Finally, if C is also $\|.\|$-complete, then it follows from theorem (3.9) that C is β-generating for each $\beta > \alpha$.

Theorem (3.9) implies that if a cone C in a normed space $(F, \|.\|)$ is $\|.\|$-complete then C is a ℬ-cone if and only if it is a strict ℬ-cone. The following proposition demonstrates another class of vector spaces with such a property.

(4.8) **PROPOSITION.** *Let $(E, C, \mathscr{P})$ be an ordered convex space and let the topological dual E' be ordered by the dual cone C'. Suppose that every $\sigma(E', E)$-bounded subset of C' is equicontinuous. Then C' is a strict ℬ-cone in $(E', \sigma(E', E))$ if (and only if) it is a ℬ-cone in $(E', \sigma(E', E))$.*

Proof. Suppose C' is a ℬ-cone in $(E', \sigma(E', E))$. Let B be a $\sigma(E', E)$-bounded subset of E'. Then there exists a $\sigma(E', E)$-bounded subset of E' such that $B \subseteq \overline{A \cap C' - A \cap C'}$. Clearly we can further assume, without loss of generality, that A is $\sigma(E', E)$-closed. Then $A \cap C'$ is $\sigma(E', E)$-closed and bounded; also, by assumption, $A \cap C'$ is equi-continuous and hence $\sigma(E', E)$-compact in view of the Alaoglu–Bourbaki theorem. Consequently $A \cap C' - A \cap C'$ must be $\sigma(E', E)$-closed, hence $B \subseteq \overline{A \cap C' - A \cap C'} = A \cap C' - A \cap C'$. This shows that C is a strict ℬ-cone in $(E', \sigma(E', E))$.

It is clear that if an ordered convex space $(E, C, \mathscr{P})$ is barrelled then every positive $\sigma(E', E)$-bounded subset of E' must be equicontinuous, and that the strong dual of a semi-reflexive space is barrelled. Thus we have the following corollary immediately.

(4.9) **COROLLARY.** *Let $(E, C, \mathscr{P})$ be an ordered convex space. If E is semi-reflexive, and if C is $\mathscr{P}$-closed, then C is a strict ℬ-cone in $(E, \mathscr{P})$ if (and only if) it is a ℬ-cone in $(E, \mathscr{P})$.*

5

LOCALLY O-CONVEX SPACES

Let (E, C) be an ordered vector space. Recall that a subset A of E is order-convex (full) if $[a_1, a_2] \subseteq A$ whenever $a_1, a_2 \in A$ and $a_1 \leqslant a_2$. A vector topology $\mathscr{P}$ in (E, C) is said to be *locally order-convex* (or *locally full*) if it admits a neighbourhood-base at 0 consisting of order-convex sets. In this case, we shall say that $(E, C, \mathscr{P})$ is a *locally order-convex space* (*locally full space*) and the cone C is called a *normal cone* in $(E, \mathscr{P})$. Assuming $\mathscr{P}$ to be Hausdorff (as we always do), a normal cone C must be a proper cone. In fact, suppose $x \in C \cap -C$. Let $\mathscr{U}$ be a $\mathscr{P}$-neighbourhood-base at 0 consisting of *order-convex* sets. Then $x \in [0, 0] \subseteq U$ for each $U \in \mathscr{U}$. Since $\mathscr{P}$ is Hausdorff, $\{0\} = \cap\{U : U \in \mathscr{U}\}$; hence $x = 0$.

(5.1) **Theorem.** *Let $(E, C, \mathscr{P})$ be an ordered topological vector space. Then the following statements are equivalent:*

(a) *$\mathscr{P}$ is locally order-convex;*

(b) *if $\{x_n, n \in D, \leqslant\}$ and $\{y_n, n \in D, \leqslant\}$ are two nets such that $0 \leqslant x_n \leqslant y_n$ for each $n \in D$ and if $\{y_n\}$ converges to 0 in $(E, \mathscr{P})$ then so does $\{x_n\}$;*

(c) *for any $\mathscr{P}$-neighbourhood W at 0, there exists a $\mathscr{P}$-neighbourhood V of 0 such that $(V - C) \cap C \subseteq W$;*

(d) *for any $\mathscr{P}$-neighbourhood W at 0 there exists a circled $\mathscr{P}$-neighbourhood U of 0 such that $[U] \subseteq W$;*

(e) *$\mathscr{P}$ admits a neighbourhood-base at 0 consisting of circled and order-convex sets.*

Proof. (a) $\Rightarrow$ (b): Straightforward.

(b) $\Rightarrow$ (c): Suppose on the contrary that there exists a $\mathscr{P}$-neighbourhood W of 0 such that

$$(V - C) \cap C \nsubseteq W$$

whenever $V \in \mathscr{U}$, where $\mathscr{U}$ denotes the family of all $\mathscr{P}$-neighbourhoods of 0. For each $V \in \mathscr{U}$, there exists $x_V \in (V - C) \cap C$ but $x_V \notin W$. Let $y_V \in V$ be such that $0 \leqslant x_V \leqslant y_V$. Notice that $\{y_V, V \in \mathscr{U}, \supseteq\}$ is a net in $(E, \mathscr{P})$ converging to 0; but $\{x_V\}$ does not converge to 0.

(c) $\Rightarrow$ (d): Take a circled $\mathscr{P}$-neighbourhood W_1 of 0 such that $W_1+W_1 \subseteq W$. By (c), there exists a $\mathscr{P}$-neighbourhood W_2 of 0 such that $(W_2-C) \cap C \subseteq W_1$. Then take a circled $\mathscr{P}$-neighbourhood W_3 of 0 such that $W_3+W_3 \subseteq W_2$. Let $U = W_1 \cap W_3$. Then U is a circled $\mathscr{P}$-neighbourhood of 0 with the property that $[U] \subseteq W$.

Finally, the implications (d) $\Rightarrow$ (e) $\Rightarrow$ (a) are trivial, and the proof of the theorem is complete.

(5.2) PROPOSITION. *Let $(E, C, \mathscr{P})$ be an ordered topological vector space. If $\mathscr{P}$ is locally ordered-convex then the order-convex hull of each $\mathscr{P}$-bounded subset of E is $\mathscr{P}$-bounded. The converse also holds in the case when $\mathscr{P}$ is metrizable.*

Proof. Let B be a $\mathscr{P}$-bounded subset of E, and V a $\mathscr{P}$-neighbourhood of 0. If $\mathscr{P}$ is locally order-convex then there exists an order-convex $\mathscr{P}$-neighbourhood U of 0 such that $U \subseteq V$. Hence B is absorbed by U and so $[B]$ is absorbed by $[U] = U$; in particular $[B]$ is absorbed by V. This shows that $[B]$ is $\mathscr{P}$-bounded.

Conversely, suppose $\mathscr{P}$ is a *metrizable* vector topology in (E, C) such that the order-convex hull of each $\mathscr{P}$-bounded subset of E is $\mathscr{P}$-bounded. We have to show that $\mathscr{P}$ is locally order-convex. If not then, in view of the preceding theorem, there exists a $\mathscr{P}$-neighbourhood W of 0 such that

$$(V-C) \cap C \nsubseteq W,$$

whenever V is a $\mathscr{P}$-neighbourhood of 0. Now take a countable neighbourhood-base $\{V_n : n = 1, 2,...\}$ at 0 such that $V_{n+1}+V_{n+1} \subseteq V_n$ for each n. Then

$$\left(\frac{1}{n} V_n - C\right) \cap C \nsubseteq W \qquad (n = 1, 2,...).$$

Hence, for each n there exist x_n in $\left(\frac{1}{n} V_n - C\right) \cap C$ and y_n in $\frac{1}{n} V_n$ such that

$$0 \leqslant x_n \leqslant y_n \in \frac{1}{n} V_n \quad \text{and} \quad x_n \notin W.$$

Since each $ny_n \in V_n$, it is clear that $\{ny_n\}$ is a sequence in $(E, \mathscr{P})$ converging to 0; hence that the set

$$A = \{0\} \cup \{ny_n : n = 1, 2,...\}$$

is $\mathscr{P}$-bounded. Notice that $nx_n \in [0, ny_n] \subseteq [A]$. Since $nx_n \notin nW$, it follows that $[A]$ is not contained in any nW ($n = 1, 2,...$). This implies that the order-convex hull $[A]$ of the $\mathscr{P}$-bounded set is not $\mathscr{P}$-bounded, which is a contradiction.

(5.3) COROLLARY. *Let $(E, C, \mathscr{P})$ be a metrizable ordered topological vector space. Then $\mathscr{P}$ is locally order-convex if and only if there exists a locally order-convex topology $\mathscr{L}$ in E for which $\mathscr{P}$ and $\mathscr{L}$ have the same topologically bounded sets.*

Next let us consider, in ordered vector spaces, vector topologies which are both locally convex and locally order-convex; such spaces will be called *locally o-convex spaces*, and the topologies are called *locally o-convex topologies*. Since a locally order-convex topology is not necessarily locally convex, the concepts of locally order-convex topologies and of locally o-convex topologies are distinct.

Recall that by an o-convex set we mean a set which is both order-convex and convex. For example, the order-convex hull of a convex (circled) set is an o-convex (circled) set. Hence a locally convex topology $\mathscr{P}$ in an ordered vector space (E, C) is locally o-convex if and only if $\mathscr{P}$ admits a neighbourhood-base at 0 consisting of o-convex (or circled o-convex) sets.

Recall also that by an absolute-o-convex set we mean a set which is both convex and absolute-order-convex, and by a positive-o-convex set we mean a set which is both convex and positive-order-convex.

(5.4) THEOREM. *Let $(E, C, \mathscr{P})$ be an ordered convex space. Then the following statements are equivalent:*

(a) *$\mathscr{P}$ is locally o-convex;*

(b) *$\mathscr{P}$ admits a neighbourhood-base at 0 consisting of circled o-convex sets;*

(c) *$\mathscr{P}$ admits a neighbourhood-base at 0 consisting of circled absolute-o-convex sets;*

(d) *$\mathscr{P}$ admits a neighbourhood-base at 0 consisting of circled positive-o-convex sets;*

(e) *the family of all $\mathscr{P}$-continuous absolute-monotone semi-norms determines the topology $\mathscr{P}$;*

(f) *the family of all $\mathscr{P}$-continuous monotone semi-norms determines the topology $\mathscr{P}$.*

Proof. The equivalence (a) $\Leftrightarrow$ (b) was noted by the remark preceding to the theorem. It is trivial that (b) $\Rightarrow$ (c) $\Rightarrow$ (d). In view of the implication (c) $\Rightarrow$ (a) of theorem (5.1), we also have (d) $\Rightarrow$ (a). Therefore the statements (a), (b), (c), and (d) are mutually equivalent. Furthermore, by proposition (1.6), (c) $\Leftrightarrow$ (e) and (d) $\Leftrightarrow$ (f). The proof of theorem (5.4) is thus completed.

(5.5) COROLLARY. *Let C be a cone in a locally convex space $(E, \mathscr{P})$, and let $\bar{C}$ be the $\mathscr{P}$-closure of C. Then $(E, C, \mathscr{P})$ is locally o-convex if and only if $(E, \bar{C}, \mathscr{P})$ is locally o-convex.*

Let α be a positive constant. A cone C in a semi-normed space $(E, \|.\|)$ is said to be α-*normal* if the following implication holds:

$$x \leqslant y \leqslant z \quad \text{in} \quad (E, C) \;\Rightarrow\; \|y\| \leqslant \alpha.\max\{\|x\|, \|z\|\}$$

(thus α must be larger than, or equal to, 1); in this case, we also say that the semi-norm $\|.\|$ is α-normal in (E, C). If Σ denotes the closed unit ball in $(E, \|.\|)$ and $F(\Sigma)$ denotes the order-convex hull of Σ in (E, C) then C is α-normal if and only if $F(\Sigma) \subseteq \alpha\Sigma$. In this case the family of the sets of the form $\{F(r\Sigma):r > 0\} = \{rF(\Sigma):r > 0\}$ is a neighbourhood-base at 0 in E with respect to the vector topology induced by $\|.\|$. Thus the first statement in the following proposition is clear.

(5.6) PROPOSITION. *Let $\|.\|$ be a semi-norm in an ordered vector space (E, C). If there exists $\alpha \geqslant 1$ such that C is α-normal in $(E, \|.\|)$ then the $\|.\|$-topology is locally o-convex. Conversely, if $\mathscr{P}$ is a semi-normable and locally o-convex topology in (E, C) then there exists a semi-norm p on E which is 1-normal and is such that $\mathscr{P}$ is precisely the vector topology induced by p.*

Proof. It remains to show the second assertion of the proposition. If $\mathscr{P}$ is locally o-convex and semi-normable, then, in view of theorem (5.4), there exists a circled and o-convex $\mathscr{P}$-neighbourhood V of 0 in E such that $\{rV:r > 0\}$ is a neighbourhood-base at 0 in $(E, \mathscr{P})$. Then the Minkowski functional of V is 1-normal (cf. proposition (1.6)) and induces the topology $\mathscr{P}$. This completes the proof.

Clearly C is α-normal if and only if it is $(\alpha+\varepsilon)$-normal for each $\varepsilon > 0$. This remark makes the following result clear.

(5.7) COROLLARY. *Let (E, C) be an ordered vector space. Then a semi-norm on E is α-normal in (E, C) for some $\alpha > 1$ if and only if there exists a semi-norm p equivalent to the given semi-norm such that p is 1-normal.*

We now turn our attention to a study of convergence in locally o-convex spaces. The following proposition is due to Bonsall (1955) and the proof given here is due to Weston (1957*a*).

(5.8) PROPOSITION. *Let $(E, C, \mathscr{P})$ be a locally o-convex space such that C is $\mathscr{P}$-closed, and let $\{x_\tau : \tau \in D\}$ be an increasing net in E. Then x_τ converges to x with respect to $\mathscr{P}$ if and only if x_τ converges to x with respect to $\sigma(E, E')$.*

Proof. The condition is clearly necessary. To prove its sufficiency we first recall, from proposition (2.1), that $x = \sup x_\tau$. Suppose that $u_\tau = x - x_\tau$ ($\tau \in D$). Then $0 \leqslant u_\tau \downarrow$ and u_τ converges to 0 with respect to $\sigma(E, E')$. By applying the strong separation theorem, 0 is contained in the $\mathscr{P}$-closed convex hull of $\{u_\tau : \tau \in D\}$; thus, for any o-convex and circled $\mathscr{P}$-neighbourhood V of 0, there exists a finite subset

$$\{\tau_i : i = 1, 2, ..., n\}$$

of D and real numbers $\lambda_i \geqslant 0$ with $\sum_{i=1}^{n} \lambda_i = 1$ such that $\sum_{i=1}^{n} \lambda_i u_{\tau_i} \in V$; since D is a direct set, there exists $\tau_0 \in D$ such that $\tau_i \leqslant \tau_0$ for all $i = 1, 2, ..., n$. We have, therefore, $0 \leqslant u_\tau = \sum_{i=1}^{n} \lambda_i u_\tau \leqslant \sum_{i=1}^{n} \lambda_i u_{\tau_i}$ whenever $\tau \geqslant \tau_0$; it follows from the order-convexity of V that $u_\tau \in V$ whenever $\tau \geqslant \tau_0$, and hence that x_τ converges to x with respect to $\mathscr{P}$.

We recall that if $(E, \mathscr{P})$ is a locally convex space and if $\mathscr{B}$ is a family of $\mathscr{P}$-bounded sets in E, we say that $\mathscr{B}$ is *saturated* if

(a) $\mathscr{B}$ contains arbitrary subsets of each of its members;

(b) $\lambda A \in \mathscr{B}$ for all $\lambda \in \mathbf{R}$ and $A \in \mathscr{B}$;

(c) for each finite subfamily $\{A_i : i = 1, 2, ..., n\}$, say, of $\mathscr{B}$ the closed circled convex hull of $\bigcup_{i=1}^{n} A_i$ is in $\mathscr{B}$.

The following theorem, due to Schaefer (1966), may be regarded as one of the central results in the theory of ordered topological vector spaces.

(5.9) THEOREM. *Let $(E, C, \mathscr{T})$ be an ordered convex space with the topological dual E', and let $\mathscr{C}$ be a saturated family consisting of $\sigma(E', E)$-bounded subsets of E' for which the linear hull of $\cup \{B : B \in \mathscr{C}\}$ is $\sigma(E', E)$-dense in E'. Suppose that $\mathscr{P}$ is the $\mathscr{C}$-topology on E.*

(a) *If C' is an $\mathscr{C}$-cone then $\mathscr{P}$ is a locally o-convex topology.*

(b) *Suppose that $\mathscr{P}$ is a locally o-convex topology. If $\mathscr{P}$ is consistent with the duality $\langle E, E' \rangle$, then C' is a strict $\mathscr{C}$-cone.*

Proof. (a) Since the linear hull of $\cup \{B : B \in \mathscr{C}\}$ is $\sigma(E', E)$-dense in E', the topology $\mathscr{P}$ is a (Hausdorff) locally convex topology in E. Let V be a $\mathscr{P}$-neighbourhood of 0 in E. Then there exists a circled convex set A in $\mathscr{C}$ such that $A^0 \subseteq V$, where the polar is taken in E. Since C' is a $\mathscr{C}$-cone, A is contained in the $\sigma(E', E)$-closure $\overline{D(B)}$ of the de-composable kernel of B for some circled convex B in $\mathscr{C}$. Then $(\overline{D(B)})^0 = (D(B))^0$ is a $\mathscr{P}$-neighbourhood of 0. Clearly $(D(B))^0$ is order-convex (cf. the proof of theorem (2.11)) and contained in A^0 and hence in V. This shows that $\mathscr{P}$ is locally o-convex.

(b) Conversely, if $\mathscr{P}$ is locally o-convex, then each $\mathscr{P}$-equicontinuous subset of E^* is contained in a decomposable equicontinuous set. If, in addition, $\mathscr{P}$ is consistent with the duality $\langle E, E' \rangle$, then $\mathscr{C}$ is a fundamental family for the equicontinuous sets. It follows in particular that each member of $\mathscr{C}$ is contained in some decomposable member of $\mathscr{C}$, i.e. C' is a strict $\mathscr{C}$-cone. This completes the proof.

The preceding theorem can be restated in the following form (cf. theorem (3.10)).

(5.10) THEOREM. *Let $(E, C, \mathscr{P})$ be an ordered convex space with the topological dual E', and let $\mathscr{C}$ be a saturated family consisting of relatively $\sigma(E', E)$-compact subsets of E'. If C' is an $\mathscr{C}$-cone then it is a strict $\mathscr{C}$-cone. In particular, if we take ξ to be a fundamental family of equicontinuous subsets of E', then $\mathscr{P}$ is a locally o-convex topology if and only if C' is an ξ-cone, and this is the case if and only if C' is a strict ξ-cone.*

It should be noted that if $(E, \mathscr{P})$ is barrelled then ξ coincides with the family of all $\sigma(E', E)$-bounded subsets of E'. If $(E, \mathscr{P})$ is infra-barrelled then ξ coincides with the family of all $\beta(E', E)$-bounded subsets of E'. Therefore the following result is a consequence of the preceding theorem.

(5.11) COROLLARY. *Let $(E, C, \mathscr{P})$ be an ordered convex space which is barrelled. Then the following statements are equivalent:*
(a) *$(E, C, \mathscr{P})$ is locally o-convex;*
(b) *C' is a strict $\mathscr{B}$-cone in $(E', \sigma(E', E))$;*
(c) *C' is a $\mathscr{B}$-cone in $(E', \sigma(E', E))$.*

If $(E, C, \mathscr{P})$ is infrabarrelled then (a), (b), (c) *are still equivalent but with* $(E', \sigma(E', E))$ *in* (b) *and* (c) *replaced by* $(E', \beta(E', E))$.

The following result should be compared with corollary (3.14).

(5.12) COROLLARY. *Let* $(E, C, \mathscr{P})$ *be an ordered convex space. Then* $E' = C' - C'$ *if and only if* $(E, C, \sigma(E, E'))$ *is locally o-convex. In particular, if* $\mathscr{P}$ *is locally o-convex then so is* $\sigma(E, E')$.

(5.13) COROLLARY. *Let* $(E, C, \mathscr{P})$ *be a metrizable ordered convex space for which* E *is of the second category, and let* C *be* $\mathscr{P}$*-complete. If* $\sigma(E', E)$ (*or* $\beta(E', E)$) *is a locally o-convex topology on* E' *then* $(E, C, \mathscr{P})$ *is locally decomposable, and hence* E *must be* $\mathscr{P}$*-complete.*

Proof. If $\beta(E', E)$ is locally o-convex then the order-convex hull in E' of each $\beta(E', E)$-bounded subset is $\beta(E', E)$-bounded. Since $\mathscr{P}$ is metrizable, $\beta(E', E)$-bounded sets are $\mathscr{P}$-equicontinuous; it follows that the order-convex hull in E' of each $\mathscr{P}$-equicontinuous subset is $\mathscr{P}$-equicontinuous. By theorem (3.11), $\mathscr{P}$ has the nearly-open decomposition property, and hence is locally decomposable since C is $\mathscr{P}$-complete (cf. theorem (3.8)).

Similarly, if $\sigma(E', E)$ is locally o-convex then, in view of the preceding corollary, C must be generating in E. It follows from theorem (3.8) that $\mathscr{P}$ is locally decomposable.

(5.14) COROLLARY. *Let* $(E, C, \mathscr{P})$ *be an ordered convex space. If* C *is an* $\mathscr{B}$*-cone in* $(E, \mathscr{P})$ *then* $(E', C', \beta(E', E))$ *is locally o-convex. The condition is also necessary for the case when* $(E, \mathscr{P})$ *is semi-reflexive and* C *is* $\mathscr{P}$*-closed.*

Proof. The first assertion follows from (a) of theorem (5.9). Conversely, if $(E, \mathscr{P})$ is semi-reflexive and C is $\mathscr{P}$-closed, then E may be regarded as the dual of $(E', C', \beta(E', E))$ with the 'dual cone' C. If, further, $(E', C', \beta(E', E))$ is locally o-convex then it follows from theorem (5.10) that the dual cone C is a strict $\mathscr{B}$-cone in $(E, \mathscr{P})$ and *a fortiori* a $\mathscr{B}$-cone.

Specializing in ordered *normed* spaces, theorem (5.9) can be strengthened as follows.

(5.15) THEOREM (Grosberg–Krein). *Let* $(E, C, \|.\|)$ *be an ordered normed space and* α *a positive real number. Then* $(E', C', \|.\|)$ *is* α*-generating if and only if* $(E, C, \|.\|)$ *is* α*-normal. This is the case if and* *only if*

$$\|x\| \leqslant \alpha.\sup\{|f(x)|:f \in C', \|f\| \leqslant 1\}, \text{ for all } x \in E.$$

Proof. Let Σ, Σ' denote the closed unit balls in E and E' respectively. By the duality theorem (theorem (2.11)) we have

$$(F(\Sigma))^{\pi} = D(\Sigma^{\pi}) = D(\Sigma').$$

Therefore $(E, C, \|.\|)$ is α-normal

$$\Leftrightarrow F(\Sigma) \subseteq \alpha\Sigma$$

$$\Leftrightarrow (F(\Sigma))^{\pi} \supseteq (\alpha\Sigma)^{\pi} = \frac{1}{\alpha} \Sigma'$$

$$\Leftrightarrow \alpha.D(\Sigma') \supseteq \Sigma'$$

$$\Leftrightarrow (E', C', \|.\|) \text{ is } \alpha\text{-generating.}$$

This proves the first assertion of the theorem; the second assertion follows easily from the first and the Hahn–Banach theorem.

The following theorem was proved by Ellis (1964), and in a somewhat weaker form by Andô (1962).

(5.16) THEOREM (Andô–Ellis). *Let $(E, C, \|.\|)$ be an ordered normed space and suppose that C is $\|.\|$-complete. Let $\alpha \geqslant 1$. Then the following statements are equivalent:*
   (a) *C' is α-normal in $(E', \|.\|)$;*
   (b) *C is nearly α-generating in $(E, \|.\|)$;*
   (c) *C is $(\alpha+\varepsilon)$-generating in $(E, \|.\|)$ for any $\varepsilon > 0$.*
If one (and hence all) of the equivalent properties is satisfied then $(E, \|.\|)$ is complete.

Proof. In view of theorem (3.9), we have only to show that (a) $\Leftrightarrow$ (b). Since C is $\|.\|$-closed, the polar $(C')^0$ of C', taken in E, is exactly $-C$ (cf. the proof of theorem (2.11)). Thus

$$(\Sigma'+C')^0 = (\Sigma')^0 \cap (C')^0 = -(\Sigma \cap C).$$

By the Alaoglu–Bourbaki theorem, $\Sigma'+C'$ is $\sigma(E', E)$-closed. In view of the bipolar theorem, we have the following chain of equivalent statements:

C' is α-normal

$\Leftrightarrow (\Sigma'+C') \cap (\Sigma'-C') \subseteq \alpha\Sigma'$

$\Leftrightarrow \{(\Sigma'+C') \cap (\Sigma'-C')\}^0 \supseteq \dfrac{1}{\alpha}\Sigma$

$\Leftrightarrow \overline{\mathrm{co}}((\Sigma'+C')^0 \cup (\Sigma'-C')^0) \supseteq \dfrac{1}{\alpha}\Sigma$

$\Leftrightarrow \overline{\mathrm{co}}(-(\Sigma \cap C) \cup (\Sigma \cap C)) \supseteq \dfrac{1}{\alpha}\Sigma$

$\Leftrightarrow \overline{D(\Sigma)} \supseteq \dfrac{1}{\alpha}\Sigma$

$\Leftrightarrow C$ is nearly α-generating,

proving that (a) $\Leftrightarrow$ (b).

(5.17) COROLLARY. *Let $(E, C, \|.\|)$ be an ordered Banach space, and suppose that C is $\|.\|$-closed. Then $\beta(E', E)$ is a locally o-convex topology on E' if and only if $\sigma(E', E)$ is a locally o-convex topology on E'.*

We conclude this section with a method constructing a locally full topology from a vector topology.

Let $(E, C, \mathscr{P})$ be an ordered topological vector space, and let $\mathscr{U}$ be a neighbourhood-base at 0 for $\mathscr{P}$. Suppose that

$$F(\mathscr{U}) = \{[V]: V \in \mathscr{U}\} = \{F(V): V \in \mathscr{U}\}.$$

It is easily seen that there exists a unique locally full (locally order-convex) topology $\mathscr{P}_{\mathrm{F}}$, say, such that $F(\mathscr{U})$ is a neighbourhood-base at 0 for $\mathscr{P}_{\mathrm{F}}$; moreover, $\mathscr{P}_{\mathrm{F}}$ is the least upper bound of all locally full (locally order-convex) topologies which are coarser than $\mathscr{P}$. This topology $\mathscr{P}_{\mathrm{F}}$ is referred to as the *locally full (locally order-convex) topology associated with $\mathscr{P}$*. If $\mathscr{P}$ is locally convex, then $\mathscr{P}_{\mathrm{F}}$ is a locally o-convex topology; in this case, $\mathscr{P}_{\mathrm{F}}$ is called the *locally o-convex topology associated with $\mathscr{P}$*. It should be noted that $\mathscr{P}_{\mathrm{F}}$ need not be Hausdorff. Observe that the $\mathscr{P}$-closure of C coincides with the $\mathscr{P}_{\mathrm{F}}$-closure of C; it then follows that $\mathscr{P}_{\mathrm{F}}$ is Hausdorff if the $\mathscr{P}$-closure of C is a *proper* cone. The following dual characterization of the topology $\mathscr{P}_{\mathrm{F}}$ should be compared with theorem (3.10).

(5.18) THEOREM. *Let $(E, C, \mathscr{P})$ be an ordered convex space, and let $\mathscr{P}_{\mathrm{F}}$ be the locally o-convex topology associated with $\mathscr{P}$. Then $\mathscr{P}_{\mathrm{F}}$ is the topology on E of uniform convergence on the decomposable $\mathscr{P}$-equicontinuous subsets of E'. Consequently $\mathscr{P}$ is locally o-convex (i.e. $\mathscr{P} = \mathscr{P}_{\mathrm{F}}$) if and only if each $\mathscr{P}$-equicontinuous subset of E' is contained in a decomposable $\mathscr{P}$-equicontinuous subset of E'.*

Proof. Let $\mathscr{U}$ be a neighbourhood base at 0 in $(E, \mathscr{P})$ consisting of circled convex sets. Let $F(\mathscr{U}) = \{F(V): V \in \mathscr{U}\}$. Then $\mathscr{P}$ is the topology of uniform convergence on $\{V^0: V \in \mathscr{U}\}$ and $\mathscr{P}_{\mathrm{F}}$ is that on

$$\{W^0: W \in F(\mathscr{U})\} = \{F(V)^0: V \in \mathscr{U}\}.$$

Thus the theorem follows immediately from theorem (2.11).

(5.19) COROLLARY. *The topological dual $(E, \mathscr{P}_{\mathrm{F}})'$ of $(E, \mathscr{P}_{\mathrm{F}})$ is equal to $C' - C'$, where C' denotes the dual cone in $E' = (E, \mathscr{P})'$. In particular, if $\mathscr{P}$ is locally o-convex then E' is decomposable in E^*.*

Remark. $C' - C'$ is the decomposable kernel of $(E, \mathscr{P})'$.

Finally we consider the permanence property of locally o-convex spaces; the following result should be compared with theorem (3.16):

(5.20) THEOREM. *Let $\{(E_\alpha, C_\alpha, \mathscr{P}_\alpha): \alpha \in \Gamma\}$ be a family of ordered convex spaces, and let $F = \bigoplus_{\alpha \in \Gamma} E_\alpha$ be the algebraic direct sum of E_α, ordered by the product cone. Let $\mathscr{P}_{\alpha, \mathrm{F}}$ be the locally o-convex topology associated with $\mathscr{P}_\alpha$ and let $(\bigoplus_{\alpha \in \Gamma} \mathscr{P}_\alpha)_{\mathrm{F}}$ be the locally o-convex topology associated with $\bigoplus_{\alpha \in \Gamma} \mathscr{P}_\alpha$. Then*

$$\left(\bigoplus_{\alpha \in \Gamma} \mathscr{P}_\alpha \right)_{\mathrm{F}} = \bigoplus_{\alpha \in \Gamma} \mathscr{P}_{\alpha, \mathrm{F}}. \tag{5.1}$$

Consequently the following assertions hold:

(a) *if each $\mathscr{P}_\alpha$ is locally o-convex then so is $\bigoplus_{\alpha \in \Gamma} \mathscr{P}_\alpha$;*

(b) *the decomposable kernel in F^* of the product $\prod_{\alpha \in \Gamma}(E_\alpha, \mathscr{P}_\alpha)'$ is precisely the product of the decomposable kernels in E_α^* of $(E_\alpha, \mathscr{P}_\alpha)'$.*

Proof. For each α in Γ, let V_α be a circled convex neighbourhood of 0 in $(E_\alpha, \mathscr{P}_\alpha)$. Let

$$A = \mathrm{co}\left(\bigcup_{\alpha \in \Gamma} V_\alpha \right) \quad \text{and} \quad B = \mathrm{co}\left(\bigcup_{\alpha \in \Gamma} F(V_\alpha) \right).$$

To prove the formula (5.1), it is sufficient to show that $B \subseteq F(A) \subseteq 3B$. It is straightforward to verify $B \subseteq F(A)$; it then remains to demonstrate that $F(A) \subseteq 3B$. Let $x \in F(A)$ and suppose that $a_1 \leqslant x \leqslant a_2$, where $a_1, a_2 \in A$. Let

$$y = \tfrac{1}{2}(x - a_1) \quad \text{and} \quad a = \tfrac{1}{2}(a_2 - a_1).$$

Then $0 \leqslant y \leqslant a \in A \subseteq B$. By the definition of B there exist a finite number of indices $\alpha_1, \alpha_2 \ldots \alpha_n$ in Γ such that

$$a = \sum_{i=1}^{n} \lambda_i b_i,$$

where each $b_i \in F(V_{\alpha_i})$ and $\lambda_i \geqslant 0$, $\sum_{i=1}^{n} \lambda_i = 1$. For each $\alpha \in \Gamma$, let π_α denote the αth projection on the αth coordinate space E_α. Then, since $0 \leqslant y \leqslant a$, it follows that

$$0 \leqslant \pi_{\alpha_i}(y) \leqslant \pi_{\alpha_i}(a) = \lambda_i b_i \in \lambda_i F(V_{\alpha_i}) \subseteq \lambda_i B \qquad (i = 1, 2, \ldots n)$$

and $\pi_\alpha(y) = 0$ for all $\alpha \in \Gamma \backslash \{\alpha_1, \alpha_2 \ldots \alpha_n\}$. Hence, since $\sum \lambda_i = 1$,

$$y = \sum_{i=1}^{n} \pi_{\alpha_i}(y) \in \lambda_1 B + \lambda_2 B + \ldots + \lambda_n B = B.$$

Consequently $x = 2y + a_1 \in 2B + B = 3B$. This shows that $F(A) \subseteq 3B$ as required. Thus formula (5.1) is verified, and assertion (a) in the theorem follows immediately. The assertion (b) follows from (a), corollary (5.19), and the following well-known result

$$\left(F, \bigoplus_\alpha \mathscr{P}_\alpha \right)' = \prod_\alpha (E_\alpha, \mathscr{P}_\alpha)'.$$

The following result should be compared with theorem (3.19) and may be regarded as a dual to the preceding theorem.

(5.21) THEOREM. *Let* $\{(E_\alpha, C_\alpha, \mathscr{P}_\alpha) : \alpha \in \Gamma\}$ *be a family of ordered convex spaces, and let E be the product space ordered by the product cone C. Let* $\prod_{\alpha \in \Gamma} \mathscr{P}_\alpha$ *be the product topology and* $(\prod_{\alpha \in \Gamma} \mathscr{P}_\alpha)_F$ *the locally o-convex topology associated with* $\prod_{\alpha \in \Gamma} \mathscr{P}_\alpha$. *Then*

$$\left(\prod_{\alpha \in \Gamma} \mathscr{P}_\alpha \right)_F = \prod_{\alpha \in \Gamma} \mathscr{P}_{\alpha.F}. \tag{5.2}$$

Consequently the following assertions hold:

(a) *if each $\mathscr{P}_\alpha$ is locally o-convex then so is* $\prod_{\alpha \in \Gamma} \mathscr{P}_\alpha$;

(b) *the decomposable kernel in E^* of the direct sum $\oplus(E_\alpha, \mathscr{P}_\alpha)'$ is precisely the direct sum of decomposable kernels in E_α^* of $(E_\alpha, \mathscr{P}_\alpha)'$.*

Proof. For each $\alpha \in \Gamma$, let V_α be a circled convex neighbourhood of 0 in $(E_\alpha, \mathscr{P}_\alpha)$, and suppose all V_α, except a finite number of α in Γ, are equal to E_α. Then formula (5.2) follows from an easily-verified fact:

$$F\left(\prod_{\alpha \in \Gamma} V_\alpha\right) = \prod_{\alpha \in \Gamma} F(V_\alpha).$$

By methods similar to those used in the preceding theorem, we can verify that assertions (a) and (b) are easy consequences of formula (5.2) and corollary (5.19).

LOCALLY SOLID SPACES

LET (E, C) be an ordered vector space such that $E = C - C$ and let V be a subset of E. Suppose that

$$S(V) = \cup\{[-v, v] : v \in V\}.$$

We recall that V is absolute-order-convex if $S(V) \subseteq V$, that V is absolutely dominated if $V \subseteq S(V)$, and that V is solid if $S(V) = V$.

Let us consider a vector topology $\mathscr{P}$ in (E, C) with a neighbourhood-base $\mathscr{U}$ at 0 consisting of circled sets. Let

$$S(\mathscr{U}) = \{S(V) : V \in \mathscr{U}\},$$

$$F(\mathscr{U}) = \{F(V) : V \in \mathscr{U}\},$$

and
$$\mathscr{U} \cap C - \mathscr{U} \cap C = \{V \cap C - V \cap C : V \in \mathscr{U}\}.$$

Let $\mathscr{P}_{S}$, $\mathscr{P}_{F}$, $\mathscr{P}_{D}$ respectively denote the vector topologies in E for which $S(\mathscr{U})$, $F(\mathscr{U})$, and $\mathscr{U} \cap C - \mathscr{U} \cap C$ are respective neighbourhood-bases at 0. The topologies $\mathscr{P}_{F}$ and $\mathscr{P}_{D}$ have been studied in detail (in Chapters 3 and 5); in particular, we know that $\mathscr{P}_{F}$ is the finest locally order-convex topology in E coarser than $\mathscr{P}$, and that $\mathscr{P}_{D}$ is the coarsest topology in E with the open decomposition property and finer than $\mathscr{P}$. Similarly one can construct topologies $\mathscr{P}_{FD} = (\mathscr{P}_{F})_{D}$, $\mathscr{P}_{DF} = (\mathscr{P}_{D})_{F}$, etc.

(6.1) THEOREM. *Let $(E, C, \mathscr{P})$ be an ordered topological vector space such that $E = C - C$. Then the following statements hold:*
(a) *$\mathscr{P}_{S}$ is locally order-convex and has the open decomposition property;*
(b) *$\mathscr{P}_{FD} = \mathscr{P}_{S} = \mathscr{P}_{DF}$.*

Proof. Note that if $V_1 + V_1 \subseteq V$ then $S(V_1) + S(V_1) \subseteq S(V)$ and $2S(V_1) \subseteq S(V) \cap C - S(V) \cap C$; hence $\mathscr{P}_{S}$ has the open decomposition property. Also, since each $S(V)$ is absolute-order-convex, $\mathscr{P}_{S}$ must be locally order-convex by theorem (5.1). Thus (a) is proved. Next we show that

$$\mathscr{P}_{F} \leqslant \mathscr{P}_{S} \leqslant \mathscr{P}_{D}. \tag{6.1}$$

The first inequality is obvious since $F(V) \supseteq S(V)$, and the second inequality is true since

$$V^* \cap C - V^* \cap C \subseteq S(V)$$

whenever

$$V^* + V^* \subseteq V.$$

By inequality (6.1) and (a) of this theorem, we immediately have

$$\mathscr{P}_{\mathrm{FD}} \leqslant \mathscr{P}_{\mathrm{S}} \leqslant \mathscr{P}_{\mathrm{DF}}.$$

It remains to show $\mathscr{P}_{\mathrm{FD}} \geqslant \mathscr{P}_{\mathrm{DF}}$. Take a $\mathscr{P}_{\mathrm{DF}}$-neighbourhood of 0 in E, say $F(V \cap C - V \cap C)$, where V is a circled $\mathscr{P}$-neighbourhood of 0 and $F(V \cap C - V \cap C)$ is the order-convex hull of $V \cap C - V \cap C$. Then

$$F(V) \cap C - F(V) \cap C \subseteq F(V \cap C - V \cap C). \tag{6.2}$$

In fact, if $x = y - z \in F(V) \cap C - F(V) \cap C$, where $y, z \in F(V) \cap C$, then there exist $u, v \in V$ such that $0 \leqslant y \leqslant u$ and $0 \leqslant z \leqslant v$. In particular, $u, -v \in V \cap C - V \cap C$ (since 0 is in $V \cap C$), and

$$x = y - z \in [-v, u] \subseteq F(V \cap C - V \cap C),$$

proving formula (6.2). This shows that $F(V \cap C - V \cap C)$ is a $\mathscr{P}_{\mathrm{FD}}$-neighbourhood of 0, and hence that $\mathscr{P}_{\mathrm{FD}} \geqslant \mathscr{P}_{\mathrm{DF}}$.

(6.2) COROLLARY. *Let* $(E, C, \mathscr{P})$ *be an ordered topological vector space. If* $\mathscr{P}$ *is locally order-convex and* C *gives an open decomposition then* $\mathscr{P} = \mathscr{P}_{\mathrm{S}}$.

Proof. The assumptions on $(E, C, \mathscr{P})$ imply that $\mathscr{P} = \mathscr{P}_{\mathrm{F}}$ and $\mathscr{P} = \mathscr{P}_{\mathrm{D}}$; hence $\mathscr{P}_{\mathrm{FD}} = \mathscr{P}_{\mathrm{D}} = \mathscr{P}$. By theorem (6.1)(b) we conclude that $\mathscr{P} = \mathscr{P}_{\mathrm{S}}$.

A locally convex topology $\mathscr{P}$ on an ordered vector space (E, C) will be called a *locally solid topology* if $\mathscr{P}$ admits a neighbourhood-base at 0 consisting of solid sets. This is the case if and only if $\mathscr{P} = \mathscr{P}_{\mathrm{S}}$. Further, since $S(V)$ is circled and convex whenever V is circled and convex, $\mathscr{P}$ is locally solid if and only if it admits a neighbourhood-base at 0 consisting of circled convex and solid sets. If $\mathscr{P}$ is locally convex, $\mathscr{P}_{\mathrm{S}}$ is called the *locally solid topology associated with* $\mathscr{P}$.

A semi-norm p on (E, C) is called a *Riesz semi-norm* if it satisfies the following two conditions:

(a) p is absolute-monotone: $-u \leqslant x \leqslant u$ in $E \Rightarrow p(x) \leqslant p(u)$;

(b) for each $x \in E$ with $p(x) < 1$ there exists $u \in E$ with $p(u) < 1$ such that $-u \leqslant x \leqslant u$.

Clearly the condition (a) is equivalent to the absolute-order-convexity of the 'open ball' $U = \{x \in E : p(x) < 1\}$, and condition (b) is equivalent to the absolutely dominated property of U; thus p is a Riesz semi-norm if and only if U is solid. This remark, together with theorem (6.1) and corollary (6.2), makes the following theorem clear.

(6.3) THEOREM. *Let* $(E, C, \mathscr{P})$ *be an ordered convex space. Then the following statements are equivalent:*
  (a) $(E, C, \mathscr{P})$ *is a locally solid space;*
  (b) $(E, C, \mathscr{P})$ *is both locally o-convex and locally decomposable;*
  (c) $\mathscr{P}$ *is determined by a family of Riesz semi-norms on* E;
  (d) $\mathscr{P} = \mathscr{P}_{\mathrm{FD}}$;
  (e) $\mathscr{P} = \mathscr{P}_{\mathrm{DF}}$.

(6.4) COROLLARY. *Let* $(E, C, \mathscr{P})$ *be a bornological locally o-convex space. Then* $(E, C, \mathscr{P})$ *is locally solid if and only if* C *is a locally strict* $\mathscr{B}$-*cone in* $(E, \mathscr{P})$.

Proof. In view of proposition (4.1), C is a locally strict $\mathscr{B}$-cone in $(E, \mathscr{P})$ if and only if $\mathscr{P}$ is locally decomposable, i.e. if and only if $\mathscr{P} = \mathscr{P}_{\mathrm{D}}$. Since $\mathscr{P}$ is locally o-convex, it follows that C is locally strict $\mathscr{B}$-cone in $(E, \mathscr{P})$ if and only if $\mathscr{P} = \mathscr{P}_{\mathrm{FD}} = \mathscr{P}_{\mathrm{S}}$, i.e. if and only if $\mathscr{P}$ is locally solid.

(6.5) COROLLARY. *Let* $(E, C, \mathscr{P})$ *be a locally o-convex space and suppose that* $E = C - C$. *Then* $\mathscr{P}_{\mathrm{S}} = \mathscr{P}_{\mathrm{D}}$, *and the topological dual* $(E, \mathscr{P}_{\mathrm{S}})'$ *of* $(E, \mathscr{P}_{\mathrm{S}})$ *is the smallest solid subspace of* E^* *generated by* $E' = (E, \mathscr{P})'$. *In particular, if* $\mathscr{P}$ *is locally solid then* $(E, \mathscr{P})'$ *is a solid subspace of* E^*.

Proof. Since $\mathscr{P}$ is locally o-convex, it follows from theorem (6.1) that $\mathscr{P}_{\mathrm{S}} = \mathscr{P}_{\mathrm{FD}} = \mathscr{P}_{\mathrm{D}}$. Also since $\mathscr{P}_{\mathrm{S}}$ is locally o-convex and locally decomposable, it follows from corollaries (3.12) and (5.19) that $(E, \mathscr{P}_{\mathrm{S}})'$ is a solid set in E^* containing E'. Furthermore, $(E, \mathscr{P}_{\mathrm{S}})' = (E, \mathscr{P}_{\mathrm{D}})'$ is the order-convex hull in E^* of E'. Consequently $(E, \mathscr{P}_{\mathrm{S}})'$ must be the smallest solid set in E^* containing E'.

Similarly we can prove the following corollary.

(6.6) COROLLARY. *Let* $(E, C, \mathscr{P})$ *be a locally decomposable space. Then* $\mathscr{P}_{\mathrm{S}} = \mathscr{P}_{\mathrm{F}}$, *and the topological dual* $(E, \mathscr{P}_{\mathrm{S}})'$ *of* $(E, \mathscr{P}_{\mathrm{S}})$ *is the* *largest solid subspace $C'-C'$ of E^* contained in $E' = (E, \mathscr{P})'$, where $C' = C^* \cap E'$.*

(6.7) COROLLARY. *Let (E, C) be an ordered vector space such that $E = C - C$, and let F be a directed subspace of E^* such that F is total over E. Then there exists a locally solid topology $\mathscr{P}$ on E such that $(E, \mathscr{P})'$ is the solid subspace of E^* generated by F. In particular, if F itself is solid in E^* then $(E, \mathscr{P})' = F$.*

Proof. In view of corollary (5.12), (E, C) equipped with the weak topology $\sigma(E, F)$ is locally o-convex. Let $\mathscr{P} = \sigma_S(E, F)$ be the locally solid topology associated with $\sigma(E, F)$; then the result follows immediately from corollary (6.5).

Similarly, on applying corollary (6.6), we have the following result.

(6.8) COROLLARY. *Let (E, C) be an ordered vector space such that $E = C - C$, and let F be an order-convex subspace of E^* such that F is total over E. Then there exists a locally solid topology $\mathscr{P}$ on E such that $(E, \mathscr{P})'$ is the largest solid subspace of E^* contained in F. In particular, if F itself is solid in E^* then $(E, \mathscr{P})' = F$.*

Proof. In view of corollary (3.14), (E, C), equipped with the Mackey topology $\tau(E, F)$, is locally decomposable. Let $\mathscr{P} = \tau_S(E, F)$ be the locally solid topology associated with $\tau(E, F)$; then the result follows immediately from corollary (6.6).

We record another consequence of theorem (6.3).

(6.9) COROLLARY. *Let $(E, C, \mathscr{P})$ be a locally o-convex space with a generating cone C, and let $\mathscr{P}_S$ be the locally solid topology on E associated with $\mathscr{P}$. Then the following assertions hold:*

   (a) *a positive subset of E is $\mathscr{P}$-bounded if and only if it is $\mathscr{P}_S$-bounded;*

   (b) *if C is a locally strict $\mathscr{B}$-cone, then a subset of E is $\mathscr{P}$-bounded if and only if it is $\mathscr{P}_S$-bounded;*

   (c) *if C is $\mathscr{P}$-closed, then a monotone increasing net $\{x_\tau\}$ converges to x in $(E, \mathscr{P})$ if and only if it does in $(E, \mathscr{P}_S)$.*

Proof. Since $\mathscr{P}$ is locally o-convex, $\mathscr{P}_S = \mathscr{P}_D$. Thus the corollary is a restatement of propositions (3.5) and (4.3).

(6.10) COROLLARY. *Let $(E, C, \mathscr{P})$ be a locally o-convex space satisfying the property that any o-convex circled subset of E which absorbs all positive $\mathscr{P}$-bounded subsets of E is a $\mathscr{P}$-neighbourhood of 0. Then $(E, C, \mathscr{P})$ is locally solid if (and only if) $E = C - C$.*

Proof. It is not hard to see that no strictly finer locally o-convex topology on E has the same positive $\mathscr{P}$-bounded subsets of E. The result then follows from theorem (6.3) and corollary (6.9).

(6.11) COROLLARY. *Let $(E, C, \|.\|)$ be an ordered Banach space for which C is $\|.\|$-closed. Let $\mathscr{P}$ be the vector topology in E induced by $\|.\|$. Then the following statements are equivalent:*

(a) *$(E, C, \mathscr{P})$ is locally solid;*

(b) *C is α-normal and β-generating in $(E, \|.\|)$ for some $\alpha \geqslant 1, \beta \geqslant 1$;*

(c) *C and C' are generating cones in E and E' respectively;*

(d) *C is β-generating in E and C' is α-generating in E' for some $\alpha \geqslant 1, \beta \geqslant 1$;*

(e) *C' is γ-normal and α-generating in $(E', \|.\|)$, for some $\gamma \geqslant 1, \alpha \geqslant 1$;*

(f) *(E', C') is locally solid with respect to the vector topology induced by the norm $\|.\|$ on E'.*

Proof. By theorem (6.3), (a) $\Leftrightarrow$ (b) and (e) $\Leftrightarrow$ (f). By theorem (3.8) and proposition (4.6), (c) $\Leftrightarrow$ (d). Finally, by the theorems of Krein–Grosberg and Andô–Ellis (theorems (5.15) and (5.16)), (b) $\Leftrightarrow$ (d) and (b) $\Leftrightarrow$ (e).

In Banach spaces (considering the metric aspect as well as the topological aspect) we have the following more satisfactory result.

(6.12) THEOREM (Davies). *Let $(E, C, \|.\|)$ be an ordered normed space, and let U be the open unit ball in $(E, \|.\|)$. Let $(E', C', \|.\|)$ be the Banach dual space with the dual cone C', and let U', Σ' respectively denote the open and closed unit balls in $(E', \|.\|)$. We consider the following statements:*

(a) *$\|.\|$ is a Riesz norm in (E, C);*

(b) *U is solid in (E, C);*

(c) *$\|.\|$ is a Riesz norm in (E', C');*

(d) *U' is solid in (E', C');*

(e) *Σ' is solid in (E', C').*

Then (a) $\Leftrightarrow$ (b) $\Rightarrow$ (c) $\Leftrightarrow$ (d) $\Leftrightarrow$ (e). *Furthermore, if E and C are $\|.\|$-complete then* (c) $\Rightarrow$ (a) *and hence all the statements* (a)–(e) *are mutually equivalent.*

Proof. It is easy to see that (a) $\Leftrightarrow$ (b), (c) $\Leftrightarrow$ (d), and (e) $\Rightarrow$ (c). We next show that (c) $\Rightarrow$ (e). Since $\|.\|$ is absolute-monotone on (E', C'), Σ' is certainly absolute-order-convex. To see that Σ' is absolutely dominated let $f \in \Sigma'$, $f \neq 0$. Then, for each positive integer n, there exist $g_n \in E'$ with $\|g_n\| < 1$ such that

$$\pm \frac{f}{\|f\| + (1/n)} \leqslant g_n$$

(by the definition of Riesz norms). Let $h_n = \{\|f\| + (1/n)\}g_n$. Then $\|h_n\| < 1 + (1/n)$ for each n. By the Alaoglu theorem, $\{h_n\}$ has a $\sigma(E', E)$-cluster point, say h. Then $\|h\| \leqslant 1$ and $\pm f \leqslant h$. This shows that Σ' is solid. Thus (c) $\Rightarrow$ (e). Therefore statements (c), (d), (e) are equivalent.

(b) $\Rightarrow$ (e): Since U is solid, $U = S(U)$. By theorem (1.17), we then have
$$U^\pi = (S(U))^\pi = S(U^\pi),$$

i.e. $\Sigma' = S(\Sigma')$. This shows that (e) holds.

We have shown that (a) $\Leftrightarrow$ (b) $\Rightarrow$ (c) $\Leftrightarrow$ (d) $\Leftrightarrow$ (e), and it remains to show (c) $\Rightarrow$ (a) under the additional assumption that E and C are $\|.\|$-complete. Accordingly we suppose that (c) holds. Then, by the established implication (b) $\Rightarrow$ (c), the norm $\|.\|$ on the second dual space E'' must be a Riesz norm. In particular, $\|.\|$ is absolute-monotone on E'' so the norm in E must also be absolute-monotone since $(E, \|.\|)$ is isometrically isomorphic to a subspace of E''. To verify that $\|.\|$ is a Riesz norm, we have only to show that U is absolutely dominated. To this end, let us consider the vector topology $\mathscr{P}$ on E induced by $\|.\|$. By (c), it follows from theorem (6.3) and corollary (6.11) that $\mathscr{P}$ is locally solid. Hence the family $\{S(\lambda U) : \lambda > 0\} = \{\lambda S(U) : \lambda > 0\}$ is a neighbourhood-base at 0 in $(E, \mathscr{P})$. Notice then that $\alpha \overline{S(U)} \subseteq \beta S(U)$ whenever $0 < \alpha < \beta$. On the other hand, since $S(\Sigma') = \Sigma'$ (by (e)), we have (cf. theorem (1.17))

$$(S(U))^\pi = S(U^\pi) = S(\Sigma') = \Sigma',$$

so that $(S(U))^{\pi\pi} = \Sigma$. It follows from the bipolar theorem that $\Sigma \subseteq \overline{S(U)}$. Hence $\Sigma \subseteq \overline{S(U)} \subseteq (1+\varepsilon)S(U)$ for each $\varepsilon > 0$. Consequently $U \subseteq S(U)$, i.e. U is absolutely dominated.

Finally we record the following permanence properties for locally solid topologies for future references.

(6.13) THEOREM.   *The following assertions hold:*
   (a) *the product of locally solid spaces is locally solid;*
   (b) *the locally convex direct sum of locally solid spaces is locally solid.*

Proof.   By locally convex sum of $\{E_\alpha : \alpha \in \Gamma\}$ we mean the algebraic direct sum of $\{E_\alpha\}$ equipped with the locally convex direct sum topology. Thus (b) follows immediately from theorems (3.16) and (5.20). Likewise, (a) follows from theorems (3.19) and (5.21).

7

THE ORDER-BOUND TOPOLOGY

SUPPOSE that (E, C) is an ordered vector space and that $\mathscr{U}$ is the family consisting of all the circled convex subsets of E each of which absorbs all order-bounded subsets of E. It is easily seen that $\mathscr{U}$ determines a locally convex topology (not necessarily Hausdorff), denoted by $\mathscr{P}_b$ on E; and this topology $\mathscr{P}_b$ is referred to as the *order-bound topology* (or *order topology*) on E. It is also obvious that the order-bound topology $\mathscr{P}_b$ on E is the finest locally convex topology on E for which every order-bounded subset of E is topologically bounded and that $(E, C, \mathscr{P}_b)' = E^b$, consequently $\mathscr{P}_b$ is Hausdorff if and only if E^b is total over E. From now on, when we consider the order-bound topology $\mathscr{P}_b$, we always assume that $\mathscr{P}_b$ is Hausdorff. It should be noted that E^b is always an order-convex subspace of E^*.

We state some elementary properties of $\mathscr{P}_b$ as follows.

(7.1) PROPOSITION. *Let (E, C) be an ordered vector space, and let $\mathscr{P}_b$ be the order-bound topology on E. Then $\mathscr{P}_b$ is a bornological topology in E and hence is the Mackey topology $\tau(E, E^b)$. Moreover, the following statements are equivalent:*
   (a) *C is a generating cone in E;*
   (b) *C is a locally strict $\mathscr{B}$-cone in $(E, \mathscr{P}_b)$;*
   (c) *$(E, C, \mathscr{P}_b)$ is a locally decomposable space.*

Proof. It is easy to verify that $(E, \mathscr{P}_b)$ is a bornological space with the topological dual E^b; hence $\mathscr{P}_b = \tau(E, E^b)$. The equivalence between (a), (b), and (c) follows from corollary (3.14) and proposition (4.1).

A sequence $\{x_n\}$ in an ordered vector space (E, C) is called *a relative uniform null-sequence* if there exists a sequence of positive numbers α_n with $\alpha_n \to \infty$, such that $\{\alpha_n x_n\}$ is an order-bounded subset of E. It is clear that $\{x_n\}$ is a relative uniform null-sequence if and only if there exist positive numbers ε_n, with $\varepsilon_n \to 0$, and an order-bounded subset B of E such that $x_n \in \varepsilon_n B$ for all n.

(7.2) LEMMA. *Let V be a circled convex subset of (E, C). Then V absorbs all order-bounded subsets of E if and only if it absorbs all relative uniform null-sequences in E.*

Proof. Let $\{x_n\}$ be a relative uniform null-sequence in E. There exist positive numbers ε_n, with $\varepsilon_n \to 0$, and an order-bounded subset B of E such that $x_n \in \varepsilon_n B$ for all n; since V absorbs B and since $\varepsilon_n \to 0$, it follows that V must absorb $\{x_n\}$. Therefore the condition is necessary. Conversely, if there exists an order-bounded subset B of E such that the assertion $B \subset n^2 V$ is false for all n, then we have $x_n \in B$ such that $x_n \notin n^2 V$. Therefore V does not absorb the relative uniform null-sequence $\{x_n/n\}$. This completes the proof.

Let $(F, \mathscr{T})$ be a locally convex space, and let T be a linear mapping of (E, C) into F. Let us say that T is *order-bounded* if it maps each order-bounded subset of E into a $\mathscr{T}$-bounded subset of F. It is clear that each order-bounded linear mapping of (E, C) into $\mathbf{R}$ is an order-bounded linear functional on E. If K is a cone in F, T is said to be *positive* if $T(C) \subseteq K$. It should be noted that if each order-bounded subset of F is $\mathscr{T}$-bounded then positive linear mappings from $(E, C, \mathscr{P})$ into $(F, K, \mathscr{T})$ are order-bounded.

The following result was proved by Wong (1972a).

(7.3) THEOREM. *Let $(E, C, \mathscr{P})$ be an ordered topological vector space, and let $\mathscr{P}_b$ be the order-bound topology on E. Then the following statements are equivalent:*

(a) *$\mathscr{P}$ is finer than $\mathscr{P}_b$;*

(b) *each circled convex subset of E which absorbs all relative uniform null-sequences in E is a $\mathscr{P}$-neighbourhood of 0;*

(c) *each order-bounded linear mapping of $(E, C, \mathscr{P})$ into any locally convex space $(F, \mathscr{T})$ is continuous.*

Proof. (a) ⇒ (b): Let V be a circled convex subset of E which absorbs all relative uniform null-sequences in E. According to lemma (7.2), V absorbs all order-bounded subsets of E, so V is a $\mathscr{P}_b$-neighbourhood of 0; consequently V is a $\mathscr{P}$-neighbourhood of 0, since $\mathscr{P}_b$ is coarser than $\mathscr{P}$.

(b) ⇒ (c): Let T be an order-bounded linear mapping of E into F and let U be a circled convex $\mathscr{T}$-neighbourhood of 0 in F. Then $T^{-1}(U)$ is a circled convex subset of E which absorbs all order-bounded subsets of E, and so by lemma (7.2) $T^{-1}(U)$ absorbs all relative uniform null-sequences in E. Therefore $T^{-1}(U)$ is a $\mathscr{P}$-neighbourhood of 0 in E, this implies that T is continuous.

(c) $\Rightarrow$ (a): Suppose that i is the identity mapping of $(E, C, \mathscr{P})$ into $(E, \mathscr{P}_b)$. Since each order-bounded subset of E is $\mathscr{P}_b$-bounded, it follows that i is order-bounded, and hence that i is continuous. Therefore $\mathscr{P}_b$ is coarser than $\mathscr{P}$.

(7.4) COROLLARY. *Let $(E, C, \mathscr{P})$ be an ordered convex space. Then the following statements are equivalent:*

(a) *$\mathscr{P}$ is the order-bound topology $\mathscr{P}_b$;*

(b) *each circled convex subset of E which absorbs all relative uniform null-sequences in E is a $\mathscr{P}$-neighbourhood of 0, and each order-bounded subset of E is $\mathscr{P}$-bounded;*

(c) *every order-bounded linear mapping of $(E, C, \mathscr{P})$ into any locally convex space $(F, \mathscr{T})$ is continuous, and each order-bounded subset of E is $\mathscr{P}$-bounded;*

(d) *$\mathscr{P}$ is the Mackey topology $\tau(E, E')$ and $E' = E^b$.*

Proof. It is clear that (a), (b), and (c) are equivalent by theorem (7.3). The equivalence of (a) and (d) was established in (a) of proposition (7.1).

(7.5) COROLLARY. *The order-bound topology on (E, C) is the finest locally convex topology for which every relative uniform null-sequence in E is convergent to 0.*

An ordered topological vector space $(E, C, \mathscr{P})$ is said to be *fundamentally σ-order-complete* if each increasing $\mathscr{P}$-Cauchy sequence in E has a supremum in E. It is clear that if C is sequentially $\mathscr{P}$-complete then $(E, C, \mathscr{P})$ must be fundamentally σ-order-complete. In the next chapter, we shall study the fundamentally σ-order-completeness in detail.

(7.6) COROLLARY. *For any bornological ordered convex space $(E, C, \mathscr{P})$, if C is a strict $\mathscr{B}$-cone in $(E, \mathscr{P})$ and if $(E, C, \mathscr{P})$ is fundamentally σ-order-complete, then $\mathscr{P}_b$ is coarser than $\mathscr{P}$ (and hence each order-bounded linear mapping of E into any locally convex space $(F, \mathscr{T})$ is continuous); in particular, all order-bounded linear functionals (and* *certainly all positive linear functionals) on E are $\mathscr{P}$-continuous. If, in addition, each order-bounded subset of E is $\mathscr{P}$-bounded then $\mathscr{P}$ is $\mathscr{P}_{\mathrm{b}}$.*

Proof. Let V be a circled convex $\mathscr{P}_{\mathrm{b}}$-neighbourhood of 0 in E. Then V absorbs all order-bounded subsets of E. We wish to show that V is a $\mathscr{P}$-neighbourhood of 0. Since $\mathscr{P}$ is bornological, it suffices to show that V absorbs every $\mathscr{P}$-bounded subset of E. Suppose, on the contrary, that there exists a $\mathscr{P}$-bounded subset B of E which is *not* absorbed by V. Since C is a strict $\mathscr{B}$-cone, for this B there exists a circled convex $\mathscr{P}$-bounded subset A of E such that $B \subseteq A \cap C{-}A \cap C$. Then the set $A \cap C$ must *not* be absorbed by V. Hence, for each positive integer n there exists $x_n \in A \cap C$ such that $x_n \notin 2^{2n}V$. The sequence $\sum\limits_{k=1}^{n} 2^{-k}x_k$ $(n \geqslant 1)$ is an increasing $\mathscr{P}$-Cauchy sequence in E, so

$$y = \sup_n \sum_{k=1}^{n} 2^{-k}x_k$$

exists in E. Notice that $0 \leqslant 2^{-n}x_n \leqslant y$ and $2^{-n}x_n \notin 2^n V$; therefore the order-interval $[0, y]$ is *not* absorbed by the $\mathscr{P}_{\mathrm{b}}$-neighbourhood V of 0, contrary to the construction of $\mathscr{P}_{\mathrm{b}}$-topology. This shows that V must be a $\mathscr{P}$-neighbourhood of 0 and hence that $\mathscr{P}_{\mathbf{b}} \leqslant \mathscr{P}$. If, in addition, each order-bounded set in E is $\mathscr{P}$-bounded then $\mathscr{P}_{\mathrm{b}} = \mathscr{P}$ by corollary (7.4).

(7.7) COROLLARY. *Let $(E, C, \mathscr{P})$ be a metrizable ordered topological vector space, and let $(F, \mathscr{T})$ be any locally convex space. Then each of the following conditions implies that each order-bounded linear mapping of E into F is continuous (in particular, all positive linear functionals on E are $\mathscr{P}$-continuous) and hence $\mathscr{P}_{\mathrm{b}}$ is certainly coarser than $\mathscr{P}$.*

(a) *C gives an open decomposition in $(E, \mathscr{P})$, and $(E, C, \mathscr{P})$ is fundamentally σ-order complete.*

(b) *C is $\mathscr{P}$-complete and generating, and E is of the second category.*

Proof. It is enough to verify (a) since (b) follows from (a) by theorem (3.8). Let $\{V_n : n \geqslant 1\}$ be a neighbourhood-base at 0 for $\mathscr{P}$ consisting of $\mathscr{P}$-closed, circled sets such that $V_{n+1}+V_{n+1} \subseteq V_n$ for all $n \geqslant 1$. Then $\{V_n \cap C{-}V_n \cap C : n \geqslant 1\}$ is a neighbourhood-base at 0 for $\mathscr{P}$ since C gives an open decomposition in $(E, \mathscr{P})$. Let V be a circled convex set in E which absorbs all relative uniform null-sequences in E.

If V is not a $\mathscr{P}$-neighbourhood of 0, then the assertion

$$V_n \cap C - V_n \cap C \subset n^4 V$$

is false for each $n \geqslant 1$. For any n, let x_n, y_n in $V_n \cap C$ be such that $x_n - y_n \notin n^4 V$. It is clear that $\left(\sum_{k=1}^{n} k^{-2} x_k\right)$ and $\left(\sum_{k=1}^{n} n^{-2} y_k\right)$ are increasing $\mathscr{P}$-Cauchy sequences in C. By the hypothesis, $\sup_{n} \sum_{k=1}^{n} k^{-2} x_k = x$ and $\sup_{n} \sum_{k=1}^{n} k^{-2} y_k = y$ exist in C. We conclude from

$$-y \leqslant n^{-2}(x_n - y_n) \leqslant x$$

that $\{n^{-3}(x_n - y_n)\}$ is a relative uniform null-sequence for which it is not absorbed by V, contradicting our assumption on V. Therefore V is a $\mathscr{P}$-neighbourhood of 0, and so $\mathscr{P}_b$ is coarser than $\mathscr{P}$ by theorem (7.3).

It will be seen that $(E, \mathscr{P})$ is complete provided that each order-bounded set in E is $\mathscr{P}$-bounded (corollary (8.11)).

(7.8) COROLLARY. *For any ordered topological vector space $(E, C, \mathscr{P})$, if C has an interior point e then $\mathscr{P}_b$ is coarser than $\mathscr{P}$ (and hence each order-bounded linear mapping of E into any locally convex space $(F, \mathscr{T})$ is continuous); in particular, all positive linear functionals on E are $\mathscr{P}$-continuous.*

Proof. The topology induced by the gauge of $[-e, e]$ is the order-bound topology $\mathscr{P}_b$. Since e is an interior point of C, there exists a circled $\mathscr{P}$-neighbourhood V of 0 such that $e + V \subseteq C$, and hence $V \subseteq [-e, e]$; this implies that $\mathscr{P}_b$ is coarser than $\mathscr{P}$. This completes the proof.

Let $(F, \mathscr{T})$ be a topological vector space. A family $\mathscr{U}$ of $\mathscr{T}$-bounded sets in E is called a $\mathscr{T}$-*determined family* if each circled convex absorbing subset of E which absorbs all members of $\mathscr{U}$ is a $\mathscr{P}$-neighbourhood of 0. A subset B of the cone C is said to be l^1-*bounded* if for each sequence $\{x_n\}$ in B and each $\{\lambda_n\} \in l^1$ with $\lambda_n \geqslant 0$, the set $\left\{\sum_{k=1}^{n} \lambda_k x_k : n \geqslant 1\right\}$ is order-bounded.

(7.9) COROLLARY. *Let $(E, C, \mathscr{P})$ be an ordered topological vector space. If there exists a $\mathscr{P}$-determined family $\mathscr{U}$ of $\mathscr{P}$-bounded subsets of E such that each member $\mathscr{U}$ is contained in the difference of two l^1-bounded subsets of C, then $\mathscr{P}_b$ is coarser than $\mathscr{P}$ (and hence each order-bounded* *linear mapping of E into any locally convex space* $(F, \mathscr{T})$ *is continuous);* *in particular, all positive linear functionals on E are* $\mathscr{P}$*-continuous.*

Proof. Let V be any circled convex set in E which absorbs all relative uniform null-sequences in E. If there exists $A \in \mathscr{U}$ such that the assertion $A \subseteq n^4 V$ is false for all $n > 1$, then the assertion $A_1 - A_2 \subseteq n^4 V$ is false for all $n > 1$, where A_1 and A_2 are l^1-bounded subsets of C such that $A \subseteq A_1 - A_2$. For any n, there exist $x_n \in A_1$, $y_n \in A_2$ such that $n^{-3}(x_n - y_n) \notin nV$. It is easily seen that

$$\{n^{-3}(x_n - y_n) : n \geqslant 1\}$$

is a relative uniform null-sequence for which it is not absorbed by V, since $\left\{ \sum_{k=1}^{n} k^{-2} x_k \right\}$ and $\left\{ \sum_{k=1}^{n} k^{-2} y_k \right\}$ are order-bounded. This contradiction shows that V must be a $\mathscr{P}$-neighbourhood of 0, and hence the result follows from theorem (7.3).

Let $(F, \mathscr{T})$ be a locally convex space. A sequence $\{y_n\}$ in F is called a *local null-sequence* if there exist positive numbers α_n, with $\alpha_n \to \infty$ such that $\{\alpha_n x_n\}$ is a $\mathscr{T}$-bounded subset of F. It is clear that $\{y_n\}$ is a local null-sequence if and only if there exist positive numbers ε_n, with $\varepsilon_n \to 0$ and a $\mathscr{T}$-bounded subset B of F such that $y_n \in \varepsilon_n B$ for all n.

(7.10) THEOREM. *Let* $(F, \mathscr{T})$ *be a locally convex space, and let T be a linear mapping of* (E, C) *into F. Then the following statements are equivalent:*

(a) T *is order-bounded;*

(b) T *maps every relative uniform null-sequence in E into a local null-sequence in F;*

(c) T *maps every relative uniform null-sequence in E into a sequence in F which converges to 0 for* $\mathscr{T}$*;*

(d) T *maps every relative uniform null-sequence in E into a* $\mathscr{T}$ *bounded subset of F.*

Proof. Let $\{x_n\}$ be a relative uniform null-sequence in E. There exist positive numbers α_n, with $\alpha_n \to \infty$, and an order-bounded subset B of E such that $\alpha_n x_n \in B$ for all n. Let T be order-bounded. Then $T(B)$ is a $\mathscr{T}$-bounded subset of F, and so $\{T(x_n)\}$ is a local null-sequence in F. This proves the implication (a) $\Rightarrow$ (b). The implications (b) $\Rightarrow$ (c) $\Rightarrow$ (d) are obvious. It remains to show (d) $\Rightarrow$ (a). Suppose that T is not order-bounded. Then there exists an order-bounded subset

B of E such that $T(B)$ is not $\mathscr{T}$-bounded. Let V be a circled convex $\mathscr{T}$-neighbourhood of 0 in F such that the assertion $T(B) \subset n^2 V$ is false for all n. For each n, let x_n in B be such that $T(x_n) \notin n^2 V$. Then we have found a relative uniform null-sequence $\{x_n/n\}$ in E such that $\{T(x_n/n)\}$ is not $\mathscr{T}$-bounded. Therefore (d) must imply (a), and the proof is complete.

The following result is an immediate consequence of theorems (7.3) and (7.10).

(7.11) COROLLARY. *Let $\mathscr{P}_b$ be the order-bound topology on (E, C), and let $(F, \mathscr{T})$ be a locally convex space. A linear mapping T of $(E, C, \mathscr{P}_b)$ into $(F, \mathscr{T})$ is continuous if and only if it is sequentially continuous, and if and only if it satisfies one of the conditions* (a)–(d) *of theorem* (7.10).

Let $(E, C, \mathscr{P})$ be an ordered convex space such that each order-bounded subset of E is $\mathscr{P}$-bounded. Then each relative uniform null-sequence $\{x_n\}$ in E is $\mathscr{P}$-bounded, and so the polar, taken in E', of the set $\{x_n\}$ is absorbing; we denote the *topology of uniform convergence on all relative uniform null-sequences* in E by $\mathscr{S}'_{r_0}$, and denote the *topology of uniform convergence on all local null-sequences* in E by $\mathscr{S}'_{c_0}$.

We now present a dual characterization of the order-bound topology as follows (cf. Wong (1972a)).

(7.12) THEOREM. *Let $(E, C, \mathscr{P})$ be an ordered convex space for which C is $\mathscr{P}$-closed and generating. Then $\mathscr{P}$ is the order-bound topology $\mathscr{P}_b$ if and only if $\mathscr{P}$ satisfies the following conditions:*
(a) *$\mathscr{P}$ is the Mackey topology $\tau(E, E')$;*
(b) *each order-bounded subset of E is $\mathscr{P}$-bounded;*
(c) *E' is $\mathscr{S}'_{r_0}$-complete.*

Proof. (i) *Necessity.* It is clear that (a) and (b) are satisfied by $\mathscr{P}$. It remains to verify that E' is $\mathscr{S}'_{r_0}$-complete. Let $\{f_\tau\}$ be a $\mathscr{S}'_{r_0}$-Cauchy net in E'. Then $\{f_\tau\}$ is a $\sigma(E', E)$-Cauchy net since $\{x/n\}$ is a relative uniform null-sequence in E for any $x \in E$, and so there exists $f \in E^*$ such that f_τ converges to f pointwise on E. Let $\{x_n\}$ be any relative uniform null-sequence in E, and let A be the set consisting of $\{x_n\}$. Then f_τ converges to f uniformly on A since f_τ is a $\mathscr{S}'_{r_0}$-Cauchy net. It is clear from theorem (7.10) that each f_τ is bounded on A, so that f is bounded on A. Thus, by theorem (7.10), $f \in E^b = E'$. Since $\{x_n\}$ was arbitrary, then f_τ converges to f for $\mathscr{S}'_{r_0}$, and so E' is $\mathscr{S}'_{r_0}$-complete.

(ii) *Sufficiency.* The condition (b) implies that $E' \subseteq E^b$. We now show that $E^b \subseteq E'$. Suppose that $\{x_n\}$ is a relative uniform null-sequence in E, and that V is the circled convex hull of $\{x_n\}$. Since E' is $\mathcal{I}'_{r_0}$-complete, by Grothendieck's completeness theorem, it is sufficient to show that the restriction of each $f \in E^b$ to the set $\bar{V}$ is $\sigma(E, E')$-continuous. In fact, since C is generating there exist positive numbers λ_n, with $\lambda_n \to 0$, and $e \in C$ such that $x_n \in \lambda_n[-e, e]$, and so $x_n \in \mu[-e, e]$ for some $\mu > 0$, because $\lambda_n \to 0$; consequently $\bar{V} \subseteq \mu[-e, e]$ since C is $\mathcal{P}$-closed. On the other hand, suppose that $E_e = \bigcup_n n[-e, e]$, and that $\|x\|$ is the gauge of $[-e, e]$ in E_e. Then $\|.\|$ is a norm on E_e, and $[-e, e]$ is the closed unit ball of $(E_e, \|.\|)$, i.e. $[-e, e] = \{x \in E_e : \|x\| \leqslant 1\}$. Observe that V is a precompact subset of $(E_e, \|.\|)$. There exists a finite subset $\{y_i; 1 \leqslant i \leqslant n\}$ of V such that $V \subseteq \bigcup_{i=1}^{n} (y_i + [-e, e])$; since $[-e, e]$ is $\sigma(E, E')$-closed, it follows from

$$\bar{V} \subseteq \overline{\bigcup_{i=1}^{n} (y_i + [-e, e])} = \bigcup_{i=1}^{n} (y_i + [-e, e])$$

that $\bar{V}$ is also a precompact subset of $(E_e, \|.\|)$. By making use of Köthe (1969, § 28,5(2)), the norm topology $\|.\|$ and the topology $\sigma(E, E')$ coincide on $\bar{V}$ because $[-e, e]$ is $\sigma(E, E')$-closed. Finally, since each $f \in E^b$ is bounded on $[-e, e]$, then the restriction of f to E_e is a continuous linear functional on $(E_e, \|.\|)$, and so the restriction of f to $\bar{V}$ is $\|.\|$-continuous and certainly $\sigma(E, E')$-continuous. This completes the proof.

(7.13) COROLLARY. *Let $(E, C, \mathcal{P})$ be an ordered convex space for which C is $\mathcal{P}$-closed and generating. Suppose that $(E, \mathcal{P})$ is bornological and that each order-bounded subset of E is $\mathcal{P}$-bounded. If each $\mathcal{I}'_{r_0}$-Cauchy net in E' is an $\mathcal{I}'_{c_0}$-Cauchy net then $\mathcal{P}$ is the order-bound topology $\mathcal{P}_b$.*

Proof. By making use of theorem (7.12), we only have to verify that E' is $\mathcal{I}'_{r_0}$-complete. Suppose that $\{f_\tau\}$ is an $\mathcal{I}'_{r_0}$-Cauchy net in E', then it is an $\mathcal{I}'_{c_0}$-Cauchy net. Since $(E, \mathcal{P})$ is bornological, it follows from Köthe (1969, § 28,5(1)) that E' is $\mathcal{I}'_{c_0}$-complete; and hence there exists f in E' such that f_τ converges to f with respect to $\mathcal{I}'_{c_0}$. On the other hand, since each order-bounded subset of E is $\mathcal{P}$-bounded, then $\mathcal{I}'_{r_0}$ is coarser than $\mathcal{I}'_{c_0}$, and so f_τ converges to f for $\mathcal{I}'_{r_0}$. This completes the proof.

Let (E, C) be an ordered vector space. We recall that E has the *Riesz decomposition property* if $[0, u+w] = [0, u]+[0, w]$ whenever u and w are in C. E is called a *weakly Riesz space* (or *weakly vector lattice*) if E has the Riesz decomposition property and $E = C-C$; and E is called a *Riesz space* (or *vector lattice*) if each pair of elements x, y of E has a *least upper bound*, written $x \vee y$, in E. It is clear that if E is a Riesz space then C is a generating cone and E has the Riesz decomposition property; but the converse is not true.

Namioka (1957) has given an example to show that the order-bound topology $\mathscr{P}_b$ on an ordered vector space need not be locally o-convex; but Schaefer (1966) has shown that if E has the Riesz decomposition property then the order-bound topology $\mathscr{P}_b$ is the finest locally o-convex topology. Therefore, by making use of corollary (6.2), proposition (7.1) (c), and Schaefer's result (mentioned in the above), the order-bound topology $\mathscr{P}_b$ on a weakly Riesz space (E, C) is locally solid. We now give a more direct and elementary proof of this result.

(7.14) THEOREM. *Let (E, C) be a weakly Riesz space, then the order-bound topology $\mathscr{P}_b$ on E is the finest locally solid topology.*

Proof. We first show that $\mathscr{P}_b$ is locally solid. Suppose that V is a circled convex $\mathscr{P}_b$-neighbourhood of 0, we wish to find a solid convex $\mathscr{P}_b$-neighbourhood U of 0 such that $U \subseteq V$. To do this, let $S = \cup \{[-u, u]: u \in V \text{ and } [0, u] \subseteq V\}$. It is clear that S is solid. If $x, y \in S$ and if $0 \leqslant \lambda \leqslant 1$, there exist $u, w \in C$, with $[0, u] \subseteq V$ and $[0, w] \subseteq V$, such that $-u \leqslant x \leqslant u$ and $-w \leqslant y \leqslant w$; and thus $-(\lambda u+(1-\lambda)w) \leqslant \lambda x+(1-\lambda)y \leqslant \lambda u+(1-\lambda)w$. By the Riesz decomposition property and the convexity of V, we have that

$$\begin{aligned}
[0, \lambda u+(1-\lambda)w] &= [0, \lambda u]+[0, (1-\lambda)w] \\
&= \lambda[0, u] + (1-\lambda)[0, w] \subseteq \lambda V+(1-\lambda)V \\
&= V,
\end{aligned}$$

and thus $\lambda x+(1-\lambda)y \in S$. Therefore S is convex. Furthermore, $S \subseteq 2V$; for if $x \in S$ then there exists $u \in C$, with $[0, u] \subseteq V$, such that $-u \leqslant x \leqslant u$. Since $0 \leqslant \dfrac{x+u}{2} \leqslant u$ and since $0 \leqslant \dfrac{u-x}{2} \leqslant u$, it follows from $x = \dfrac{x+u}{2} - \dfrac{u-x}{2}$ that $x \in 2V$, and hence that $S \subseteq 2V$. On the other hand, we also have that $S \cap C = \{u \in C: [0, u] \subseteq V\}$. Indeed, obviously $\{u \in C: [0, u] \subseteq V\} \subseteq S \cap C$. Suppose that $x \in S \cap C$. Then there exists $u \in C$ with $[0, u] \subseteq V$, such that $-u \leqslant x \leqslant u$, and so $[0, x] \subseteq [0, u] \subseteq V$, therefore $x \in \{u \in C . [0, u] \subseteq V\}$; this shows that $S \cap C = \{u \in C : [0, u] \subseteq V\}$. Finally we show that S absorbs all elements in C; from this S absorbs all order-bounded subsets of E because C is generating. Suppose not; then there exists $x \in C$ such that $x \notin nS$ for all n, it then follows from $\{u \in C : [0, u] \subseteq V\} = S \cap C$ that $[0, x] \subseteq nV$ is false for all n; this gives a contradiction because V absorbs all order-bounded subsets of E. Therefore S is a solid convex $\mathscr{P}_b$-neighbourhood of 0 for which $S \subseteq 2V$, and so $U = \frac{1}{2}S$ has the desired property.

Since each order-bounded subset of E is bounded with respect to any locally solid topology on E, it follows that $\mathscr{P}_b$ is the finest locally solid topology. This completes the proof.

It is known from proposition (7.1) that $(E, C, \mathscr{P}_b)$ is bornological, where $\mathscr{P}_b$ is the order-bound topology on E. We conclude this chapter with an example which will show that the topology on a bornological ordered convex space $(E, C, \mathscr{P})$, for which every order-bounded subset of E is $\mathscr{P}$-bounded, need not be the order-bound topology.

(7.15) EXAMPLE. Let E be the vector subspace of ℓ^∞ consisting of all elements $x = (x_n : n \in \mathbf{N})$ satisfying the condition $x_n = 0$ for all but a finite number of indices n, equipped with the norm

$$\|x\| = \max\{|x_n| : n \in \mathbf{N}\}.$$

E has a natural cone C defined by

$$C = \{x = (x_n : x \in \mathbf{N}) : x_n \geqslant 0 \text{ for all } n \text{ in } \mathbf{N}\}.$$

Let
$$V = \left\{x = (x_n : x \in \mathbf{N}) : |x_n| \leqslant \frac{1}{n} \text{ for all } n \text{ in } \mathbf{N}\right\}.$$

Then V is a convex solid absorbing subset of E which is not a $\|.\|$-neighbourhood of 0. Therefore the norm topology $\|.\|$ is not the order-bound topology.

8

METRIZABLE ORDERED
TOPOLOGICAL VECTOR SPACES

THIS chapter is devoted to a study of relations between order completeness and topological completeness in metrizable ordered topological
vector spaces. We shall see that there are useful and elegant results that
do not hold in non-metrizable spaces.

Throughout this book ℓ^1 denotes the ordered Banach space consisting of all absolutely summable sequences of real numbers, equipped
with its usual norm and ordering; therefore (λ_n) in ℓ^1 is positive if and
only if each λ_n is non-negative.

(8.1) DEFINITIONS. Suppose that $(E, C, \mathscr{P})$ is an ordered topological
vector space.

(1) $(E, \mathscr{P})$ is said to be *boundedly σ-order-complete* if each sequence
in E which is increasing and $\mathscr{P}$-bounded has a supremum in E.
$(E, \mathscr{P})$ is said to be *boundedly order-complete* if each net in E which is
increasing and $\mathscr{P}$-bounded has a supremum in E.

(2) $(E, \mathscr{P})$ is said to be *fundamentally σ-order-complete* if each increasing $\mathscr{P}$-Cauchy sequence has a supremum in E.

(3) $(E, \mathscr{P})$ is said to be *monotonically sequentially complete* if each
increasing $\mathscr{P}$-Cauchy sequence is convergent in E.

(4) $(E, \mathscr{P})$ is said to be *ℓ^1-order-summable* if for each positive $\mathscr{P}$-
bounded sequence $\{u_n\}$ in E and any positive element (λ_n) in ℓ^1, the
sequences of partial sums of $\{\lambda_n u_n : n \geqslant 1\}$ have a supremum in E,
i.e. $\sup\left\{\sum_{k=1}^{n} \lambda_k u_n : n \geqslant 1\right\}$ exists in E; the supremum will be conveniently denoted by $(0) - \sum_k \lambda_k u_k$.

It is clear that an ordered topological vector space is fundamentally
σ-order-complete if and only if each *positive*, increasing $\mathscr{P}$-Cauchy
sequence has a supremum, and that if E and C are $\mathscr{P}$-complete then
$(E, \mathscr{P})$ is fundamentally σ-order-complete.

In metrizable and fundamentally σ-order-complete spaces, we have
the following useful result characterizing the local order-convexity.

(8.2) THEOREM. *For any metrizable ordered topological vector space* $(E, C, \mathscr{P})$, *if* $(E, \mathscr{P})$ *is fundamentally σ-order-complete (in particular, E and C are $\mathscr{P}$-complete) then* $(E, C, \mathscr{P})$ *is locally full if and only if each order-bounded subset of E is $\mathscr{P}$-bounded.*

Proof. The necessity is obvious. To prove the sufficiency, let $\{V_n : n \geqslant 1\}$ be a neighbourhood-base at 0 for $\mathscr{P}$ consisting of circled sets such that $V_{n+1} + V_{n+1} \subset V_n$. Suppose, on the contrary, that $(E, C, \mathscr{P})$ is not locally full; then there exists a $\mathscr{P}$-neighbourhood W of 0 such that
$$(V_n - C) \cap C \nsubseteq 2^n W \quad \text{for all } n \geqslant 1.$$
For each $n \geqslant 1$, there exist $x_n \in E$, $y_n \in V_n$ such that
$$0 \leqslant x_n \leqslant y_n \quad \text{and} \quad x_n \notin 2^n W.$$
It is clear that the sequence of the partial sums of $\{y_n\}$ is an increasing $\mathscr{P}$-Cauchy sequence. By the fundamental σ-order-completeness,
$$y = \sup\left\{\sum_{k=1}^{m} y_k : m \geqslant 1\right\} \text{ exists in } E, \text{ and so}$$
$$0 \leqslant x_n \leqslant y_n \leqslant y.$$
Since $x_n \notin 2^n W$, it follows that the order-interval $[0, y]$ is not $\mathscr{P}$-bounded. This completes the proof.

(8.3) COROLLARY. *Let* $(E, C, \mathscr{P})$ *be a metrizable, locally decomposable space, and let* $(E, \mathscr{P})$ *be fundamentally σ-order-complete. Then the following statements are equivalent:*
  (a) $(E, C, \mathscr{P})$ *is a locally o-convex space;*
  (b) $\mathscr{P}$ *is the order-bound topology* $\mathscr{P}_{\mathrm{b}}$;
  (c) *each order-bounded subset of E is $\mathscr{P}$-bounded.*
  Furthermore, if $(E, C, \mathscr{P})$ *satisfies one of* (a), (b), *and* (c) *then* $(E, C, \mathscr{P})$ *is a locally solid space; if, in addition,* (E, C) *has the Riesz decomposition property, then* $E' = E^{\mathrm{b}} = E^{\#}$ *and* E' *is a vector lattice.*

Proof. The implication (b) $\Rightarrow$ (c) is clear, and the implication (c) $\Rightarrow$ (a) follows from theorem (8.2). It remains to verify that (a) implies (b). Let $\{V_n : n \geqslant 1\}$ be a neighbourhood-base at 0 for $\mathscr{P}$ consisting of circled convex sets such that $V_{n+1} + V_{n+1} \subset V_n$. Since $(E, C, \mathscr{P})$ is locally solid, for each n there exists a solid convex $\mathscr{P}$-neighbourhood U_n of 0 such that
$$U_n \subset V_n.$$

If $\mathscr{P}$ is not $\mathscr{P}_b$, there exists a circled convex $\mathscr{P}_b$-neighbourhood W of 0 such that
$$U_n \not\subseteq 2^n W \quad \text{for all } n \geqslant 1.$$

For each $n \geqslant 1$, there exist $x_n \in E$ and $y_n \in U_n$ such that
$$-y_n \leqslant x_n \leqslant y_n \quad \text{and} \quad x_n \notin 2^n W.$$

Since $y_n \in V_n$, it follows from the fundamental σ-order-completeness that $y = \sup\left\{\sum_{k=1}^{m} y_k : m \geqslant 1\right\}$ exists in E; hence
$$-y \leqslant -y_n \leqslant x_n \leqslant y_n \leqslant y.$$

Therefore W does not absorb the order-interval $[-y, y]$ which gives a contradiction. The final conclusion that E' is a vector lattice follows from theorem (1.10).

(8.4) COROLLARY. *For any ordered Fréchet space $(E, C, \mathscr{P})$, if C is $\mathscr{P}$-closed, then $(E, C, \mathscr{P})$ is locally o-convex if and only if $E' \subset E^b$. In this case, $\mathscr{P}$ is the order-bound topology $\mathscr{P}_b$.*

Proof. Application of theorem (8.2) and corollary (8.3).

(8.5) PROPOSITION. *Let $(E, C, \mathscr{P})$ be a metrizable ordered topological vector space, and let C give an open decomposition in $(E, \mathscr{P})$. If (E, C) is Archimedean and if $(E, \mathscr{P})$ is fundamentally σ-order-complete, then C is $\mathscr{P}$-closed.*

Proof. Let $\{V_n : n \geqslant 1\}$ be a neighbourhood-base at 0 for $\mathscr{P}$ consisting of circled sets such that $V_{n+1} + V_{n+1} \subset V_n$, and let x be in the $\mathscr{P}$-closure of C. For each n, there exists $x_n \in C$ such that
$$x_n - x \in \frac{1}{n}(V_n \cap C - V_n \cap C),$$

and so there exists $u_n \in V_n \cap C$ such that $n(x_n - x) \leqslant u_n$. It is clear that $\left\{\sum_{k=1}^{n} u_k : n \geqslant 1\right\}$ is an increasing $\mathscr{P}$-Cauchy sequence. By the fundamental σ-order-completeness, $u = \sup\left\{\sum_{k=1}^{n} u_k : n \geqslant 1\right\}$ exists in E. We conclude from the Archimedean property of E and from
$$-x \leqslant x_n - x \leqslant \frac{1}{n} u,$$

that $-x \leqslant 0$ and hence that $x \in C$. This shows that C is $\mathscr{P}$-closed.

(8.6) LEMMA. *For any ordered topological vector space* $(E, C, \mathscr{P})$, *consider the following statements:*

(a) $(E, \mathscr{P})$ *is sequentially complete;*

(b) $(E, \mathscr{P})$ *is monotonically sequentially complete;*

(c) $(E, \mathscr{P})$ *is ℓ^1-order-summable;*

(d) $(E, \mathscr{P})$ *is fundamentally σ-order-complete;*

(e) $(E, \mathscr{P})$ *is boundedly σ-order-complete.*

Then the following statements hold:

(1) (a) $\Rightarrow$ (b) *and* (e) $\Rightarrow$ (d);

(2) (b) $\Rightarrow$ (c) *provided that* $(E, \mathscr{P})$ *is locally convex and C is $\mathscr{P}$-closed;*

(3) (b) $\Rightarrow$ (d) *provided that C is $\mathscr{P}$-closed;*

(4) (d) $\Rightarrow$ (c) *provided that* $(E, \mathscr{P})$ *is locally convex;*

(5) (c) $\Rightarrow$ (d) *provided that* $(E, \mathscr{P})$ *is metrizable—consequently, for a metrizable ordered convex space* (c) *and* (d) *are equivalent.*

Proof. We have only to verify assertion (5) since other assertions are obvious. Let $\{V_n : n = 1, 2, \ldots\}$ be a neighbourhood-base at 0 for $\mathscr{P}$ consisting of circled sets such that $V_{n+1} + V_{n+1} \subseteq V_n$ for all $n \geqslant 1$, and let $\{x_n\}$ be an increasing $\mathscr{P}$-Cauchy sequence. There exists a subsequence $\{x_{n_k}\}$ of $\{x_n\}$ such that

$$x_{n_{k+1}} - x_{n_k} \in 2^{-k} V_k \qquad (k \geqslant 1),$$

then $\{2^k(x_{n_{k+1}} - x_{n_k}) : k \geqslant 1\}$ is a $\mathscr{P}$-bounded sequence in C, and so, by the ℓ^1-order-summability,

$$x = \sup\left\{ \sum_{k=1}^{j} 2^{-k} 2^k (x_{n_{k+1}} - x_{n_k}) : j \geqslant 1 \right\}$$

exists in E. Since $\sum_{k=1}^{j} (x_{n_{k+1}} - x_{n_k}) = x_{n_{j+1}} - x_{n_1}$, it follows that

$$y = x + x_{n_1} = \sup\{x_{n_{j+1}} : j \geqslant 1\}$$

exists in E. We now claim that $y = \sup\{x_m : m \geqslant 1\}$; it is equivalent to verify that

$$x_m \leqslant y \quad \text{for all} \quad m.$$

In fact, for any m there exists a positive integer q such that $n_k \geqslant m$ whenever $k > q$; since $\{x_n\}$ is increasing, it follows that

$$x_m \leqslant x_{n_k} \leqslant y.$$

This completes the proof.

There are ordered Banach spaces which are fundamentally σ-order-complete but not boundedly σ-order-complete as shown by the following example.

(8.7) EXAMPLE. Consider the ordered Banach space $C[0, 1]$ consisting of all real-valued continuous functions on $[0, 1]$, equipped with its usual norm and ordering. Suppose that

$$x_n(t) = \begin{cases} 0 & \text{if } 0 \leqslant t \leqslant \frac{1}{2} \\ n(t-\frac{1}{2}) & \text{if } \frac{1}{2} \leqslant t \leqslant \frac{1}{2}+\frac{1}{n} \\ 1 & \text{if } \frac{1}{2}+\frac{1}{n} \leqslant t \leqslant 1. \end{cases}$$

Then $\{x_n\}$ is an increasing norm-bounded sequence in $C[0, 1]$ but has no supremum in $C[0, 1]$. Thus $C[0, 1]$ is not boundedly σ-order-complete. We observe that the cone

$$C = \{x \in C[0, 1]: x(t) \geqslant 0 \quad \text{for all } t \in [0, 1]\}$$

is (norm) closed. Therefore $C[0, 1]$ is fundamentally σ-order-complete.

The study of the relationship between order-completeness and topological completeness can be broken down into two stages: in the first stage we establish the fact that monotonically sequential completeness, under certain conditions, implies completeness; and in the second stage we establish some sufficient conditions to ensure that order-completeness implies monotonically sequential completeness. First, we prove the following theorem due to Jameson (1970).

(8.8) THEOREM. *For any metrizable ordered topological vector space* $(E, C, \mathscr{P})$, *if* $(E, \mathscr{P})$ *is monotonically sequentially complete and if* C *gives an open decomposition in* $(E, \mathscr{P})$, *then* $(E, \mathscr{P})$ *is complete. The converse is also true provided that* C *is* $\mathscr{P}$-*complete and generating.*

Proof. Let $\{V_n : n \geqslant 1\}$ be a neighbourhood-base at 0 for $\mathscr{P}$ consisting of circled $\mathscr{P}$-closed subsets of E for which $V_{n+1}+V_{n+1} \subseteq V_n$ for all $n \geqslant 1$, and let $B_n = V_n \cap C - V_n \cap C$ $(n \geqslant 1)$. Then B_n is a $\mathscr{P}$-neighbourhood of 0. Any $\mathscr{P}$-Cauchy sequence has a subsequence $\{z_n\}$ such that

$$z_{n+1}-z_n \in B_n \quad (n \geqslant 1).$$

There exist u_n, w_n in $V_n \cap C$ such that

$$z_{n+1}-z_n = u_n-w_n \quad (n \geqslant 1).$$

Since $z_{n+1}-z_1 = \sum_{k=1}^{n} u_k - \sum_{k=1}^{n} w_k$, then the convergence of $\{z_n\}$ is equivalent to the convergence of the sequences $\left\{\sum_{k=1}^{n} u_k : n \geqslant 1\right\}$ and $\left\{\sum_{k=1}^{n} w_k : n \geqslant 1\right\}$. Note that

$$\sum_{k=1}^{n+q} u_k - \sum_{k=1}^{n} u_k \in V_{n+1} \cap C + \ldots + V_{n+q} \cap C \subseteq V_n \cap C \subseteq V_n$$

for all q; thus $\left\{\sum_{k=1}^{n} u_k : n \geqslant 1\right\}$ is an increasing $\mathscr{P}$-Cauchy sequence, and hence converges for $\mathscr{P}$. Similarly we can show that $\left\{\sum_{k=1}^{n} w_k : n \geqslant 1\right\}$ converges for $\mathscr{P}$. Therefore E is $\mathscr{P}$-complete. The converse follows from theorem (3.8). This completes the proof.

As the second stage in establishing the relationship between order completeness and topological completeness, we prove the following result of Wong (1972b) which should be compared with the equivalence of (a) and (c) in theorem (3.8).

(8.9) THEOREM. *For any metrizable ordered topological vector space* $(E, C, \mathscr{P})$, *if* $\mathscr{P} \leqslant \mathscr{P}_{\mathrm{b}}$, *$C$ is generating, and if $(E, \mathscr{P})$ is fundamentally σ-order-complete then it is monotonically sequentially complete and locally full. If, in addition, E is of the second category, then $(E, \mathscr{P})$ is complete and C gives an open decomposition in $(E, \mathscr{P})$.*

Proof. The locally full property of $(E, C, \mathscr{P})$ is an immediate consequence of theorem (8.2). Let $\{V_n : n \geqslant 1\}$ be a neighbourhood-base at 0 for $\mathscr{P}$ consisting of circled $\mathscr{P}$-closed sets such that $V_{n+1} + V_{n+1} \subseteq V_n$ ($n \geqslant 1$). Then $\{V_n \cap C - V_n \cap C : n \geqslant 1\}$ forms a neighbourhood-base at 0 for the metrizable topology $\mathscr{P}_{\mathrm{D}}$. Notice that $\mathscr{P} \leqslant \mathscr{P}_{\mathrm{D}} \leqslant \mathscr{P}_{\mathrm{b}}$ and each increasing $\mathscr{P}$-Cauchy sequence in E is also $\mathscr{P}_{\mathrm{D}}$-Cauchy. Hence, to show that $(E, \mathscr{P})$ is monotonically sequentially complete it is sufficient to show that $(E, \mathscr{P}_{\mathrm{D}})$ is complete. Any $\mathscr{P}_{\mathrm{D}}$-Cauchy sequence in E has a subsequence $\{x_n\}$ such that

$$x_{n+1} - x_n \in n^{-2}(V_n \cap C - V_n \cap C),$$

and so there exist y_n, z_n in $V_n \cap C$ such that

$$n^2(x_{n+1} - x_n) = y_n - z_n.$$

It is easily seen that $\left\{\sum_{k=1}^{m} k^{-1}y_k : m \geqslant 1\right\}$, $\left\{\sum_{k=1}^{m} k^{-1}z_k : m \geqslant 1\right\}$ $\left\{\sum_{k=1}^{m} k^{-2}y_k : m \geqslant 1\right\}$, and $\left\{\sum_{k=1}^{m} k^{-2}z_k : m \geqslant 1\right\}$ are increasing $\mathscr{P}$-Cauchy sequences. Since $(E, \mathscr{P})$ is fundamentally σ-order-complete, it follows that

$$y_0 = (0) - \sum_k k^{-1} y_k, \qquad z_0 = (0) - \sum_k k^{-1} z_k$$

$$y = (0) - \sum_k k^{-2} y_k, \qquad z = (0) - \sum_k k^{-2} z_k$$

exist in E. We claim that $\left\{ \sum_{k=1}^m k^{-2} y_k : m \geqslant 1 \right\}$ converges to y with respect to $\mathscr{P}_D$. For each m, we have

$$0 \leqslant y - \sum_k^m k^{-2} y_k = (0) - \sum_{k=m+1}^{\infty} k^{-2} y_k \leqslant m^{-1} \left\{ (0) - \sum_{j=1}^{\infty} \frac{1}{m+j} \, y_{m+j} \right\} \leqslant m^{-1} y_0.$$

Since $\lfloor 0, y_0 \rfloor$ is $\mathscr{P}_D$-bounded, then the sequence $\left\{ \sum_{k=1}^m k^{-2} y_k : m \geqslant 1 \right\}$ converges to y with respect to $\mathscr{P}_D$. Similarly, the sequence $\left\{ \sum_{k=1}^m k^{-2} z_k : m \geqslant 1 \right\}$ converges to z with respect to $\mathscr{P}_D$. Therefore, the sequence $\{x_m\}$ converges to $y - z + x_1$ with respect to $\mathscr{P}_D$, and hence $(E, \mathscr{P}_D)$ is complete.

Finally, since $\mathscr{P}$ is coarser than $\mathscr{P}_D$, then the identity mapping from $(E, \mathscr{P}_D)$ onto $(E, \mathscr{P})$ is continuous, and hence $\mathscr{P}$ coincides with $\mathscr{P}_D$ by the open mapping theorem provided that $(E, \mathscr{P})$ is of the second category; consequently $(E, \mathscr{P})$ is complete, and C gives an open decomposition in $(E, \mathscr{P})$. This completes the proof.

Remark. It is worth to note that under the assumption of theorem (8.9), the metrizable topological vector space $(E, \mathscr{P}_D)$ has the following properties:

(a) each order-bounded subset of E is $\mathscr{P}_D$-bounded;

(b) C gives an open decomposition in $(E, \mathscr{P}_D)$;

(c) a monotone sequence in E is $\mathscr{P}_D$-Cauchy if and only if it is $\mathscr{P}$-Cauchy;

(d) $(E, \mathscr{P}_D)$ is complete.

(8.10) COROLLARY. *Let $(E, C, \mathscr{P})$ be a metrizable ordered convex space such that C is $\mathscr{P}$-closed and generating, and let $\mathscr{P}$ be coarser than $\mathscr{P}_b$. If $(E, \mathscr{P})$ is fundamentally σ-order-complete then $\mathscr{P}_b$ is the locally decomposable topology $\mathscr{P}_D$ on E associated with $\mathscr{P}$, and $(E, C, \mathscr{P}_b)$ is a complete, metrizable locally solid space.*

Proof. In view of the remark of theorem (8.9), $(E, C, \mathscr{P}_D)$ is a complete, metrizable locally decomposable space and $\mathscr{P}_D$ is coarser than

$\mathscr{P}_b$; hence, by making use of corollary (8.3), $\mathscr{P}_D = \mathscr{P}_b$ and $(E, C, \mathscr{P}_D)$ is also locally o-convex; consequently $(E, C, \mathscr{P}_b)$ is a locally solid space. This completes the proof.

(8.11) COROLLARY. *For any metrizable ordered topological vector space* $(E, C, \mathscr{P})$, *if* $\mathscr{P}$ *is coarser than* $\mathscr{P}_b$, *then each of the following conditions implies the completeness of* $(E, \mathscr{P})$:

(a) *C gives an open decomposition in* $(E, \mathscr{P})$ *and* $(E, C, \mathscr{P})$ *is fundamentally σ-order-complete;*

(b) *C is $\mathscr{P}$-complete, generating, and E is of the second category.*

If, in addition, $\mathscr{P}$ *is locally convex then* $\mathscr{P} = \mathscr{P}_b$ *and* $(E, C, \mathscr{P})$ *is a complete metrizable locally solid space.*

Proof. Since, by making use of theorem (3.8), (b) is an immediate consequence of (a), therefore we have only to show (a). According to theorem (8.9), $(E, C, \mathscr{P})$ is monotonically sequentially complete, and hence, from theorem (8.8), $(E, \mathscr{P})$ is complete since C gives an open decomposition in $(E, \mathscr{P})$. The final assertion follows from corollary (7.7).

(8.12) COROLLARY. *Let* $(E, C, \mathscr{P})$ *be a metrizable ordered convex space for which C is a locally strict $\mathscr{B}$-cone in* $(E, \mathscr{P})$, *and let* $\mathscr{P}$ *be coarser than* $\mathscr{P}_b$. *If* $(E, \mathscr{P})$ *is fundamentally σ-order-complete, then* $\mathscr{P} = \mathscr{P}_b$, *and* $(E, C, \mathscr{P})$ *is a complete metrizable locally solid space.*

Proof. Since metrizable convex spaces are bornological, it follows from proposition (4.1) that $(E, C, \mathscr{P})$ is a locally decomposable space, and hence the result follows from corollary (8.11).

9

ORDERED NORMED VECTOR SPACES

By an *ordered normed space* we mean a normed space $(E, \|.\|)$ equipped with a partial ordering $\leqslant$ induced by a cone C. Recall that $(E, C, \|.\|)$ is an approximate order-unit normed space if there exists an approximate order-unit $\{e_\lambda, \lambda \in \Lambda, \leqslant\}$ in C such that the given norm $\|.\|$ is the gauge of the circled convex set

$$S_\Lambda = \cup\{[-e_\lambda, e_\lambda]\colon \lambda \in \Lambda\}.$$

Since S_Λ is solid, the induced norm must be a Riesz norm and hence $(E, C, \|.\|)$ must be locally solid under the vector topology induced by $\|.\|$. The following result characterizes approximate order-unit normed spaces among Riesz normed spaces.

(9.1) PROPOSITION. *Let $(E, C, \|.\|)$ be an ordered normed space. Then the following statements are equivalent:*

   (a) *E is an approximate order-unit normed space;*

   (b) *C is 1-normal in $(E, \|.\|)$ and the open unit ball*

$$U = \{x \in E\colon \|x\| < 1\}$$

is directed upwards;

   (c) *U is order-convex and directed upwards;*

   (d) *U is absolute-order-convex and directed upwards;*

   (e) *U is solid and directed upwards;*

   (f) *$\|.\|$ is a Riesz norm on (E, C) and U is directed upwards.*

Proof. In view of proposition (1.6) and theorem (6.3), (b) $\Leftrightarrow$ (c) and (e) $\Leftrightarrow$ (f). If U is directed upwards then *a fortiori* it is absolutely dominated; hence (d) $\Leftrightarrow$ (e). Further, it is trivial that (c) $\Rightarrow$ (d). Thus to complete the proof we have only to show (a) $\Rightarrow$ (b) and (e) $\Rightarrow$ (a).

(a) $\Rightarrow$ (b): Let $\|.\|$ be the gauge of $S_\Lambda = \cup\{[-e_\lambda, e_\lambda]\colon \lambda \in \Lambda\}$, and suppose that $x \leqslant y \leqslant z$ in E. Let $M = \max\{\|x\|, \|z\|\}$. Let ε, with $\varepsilon > 0$, be given. Then there exist λ_x, λ_z in Λ such that

$$-(M+\varepsilon)e_{\lambda_x} \leqslant x \leqslant (M+\varepsilon)e_{\lambda_x}$$

and

$$-(M+\varepsilon)e_{\lambda_z} \leqslant z \leqslant (M+\varepsilon)e_{\lambda_z}.$$

Let $\mu \geqslant \lambda_x$, λ_z in Λ. Then we have that

$$-(M+\varepsilon)e_\mu \leqslant x \quad \text{and} \quad z \leqslant (M+\varepsilon)e_\mu.$$

Hence, since $x \leqslant y \leqslant z$,

$$-(M+\varepsilon)e_\mu \leqslant x \leqslant y \leqslant z \leqslant (M+\varepsilon)e_\mu.$$

It follows that $\|y\| \leqslant M+\varepsilon$. Since ε is arbitrary,

$$\|y\| \leqslant M = \max\{\|x\|, \|z\|\}.$$

This shows that C is 1-normal in $(E, \|.\|)$. To show that U is directed upwards, let u_1, $u_2 \in U$ and α a positive real number such that $\max\{\|u_1\|, \|u_2\|\} < 1-\alpha$. Then there exist λ_1, λ_2 in Λ such that

$$-(1-\alpha)e_{\lambda_i} \leqslant u_i \leqslant (1-\alpha)e_{\lambda_i} \qquad (i = 1, 2).$$

Let $\lambda \in \Lambda$ be such that $\lambda \geqslant \lambda_1$, λ_2 and $u = (1-\alpha)e_\lambda$. Then $u \in U$ and $u \geqslant u_1, u_2$.

(e) $\Rightarrow$ (a): Suppose the open unit ball U in $(E, C, \|.\|)$ is solid and directed upwards. Then $U \cap C$ is also directed. For each $u \in U \cap C$, let $e_u = u$. Then $\{e_u, u \in U \cap C, \leqslant\}$ is a net in C. Let

$$S = \cup\{[-e_u, e_u] : u \in U \cap C\}.$$

Since U is solid, it is not difficult to show that $S = U$. Hence the given norm $\|.\|$ is precisely the gauge of S; that is, $\|.\|$ is an approximate order-unit norm with the approximate order-unit $\{e_u\}$.

In what follows the open unit ball in a normed space $(E, \|.\|)$ is always denoted by U, i.e. $U = \{x \in E : \|x\| < 1\}$; if $(E', \|.\|)$ is the Banach dual space of $(E, \|.\|)$, then U' will denote the open unit ball in $(E', \|.\|)$. The closed unit ball in $(E, \|.\|)$ is denoted by Σ, i.e. $\Sigma = \{x \in E : \|x\| \leqslant 1\}$; Σ' will denote the closed unit ball in $(E', \|.\|)$.

An ordered normed space $(E, C, \|.\|)$ is called an *order-unit normed space* if there exists an order-unit e such that $\|.\|$ is the gauge of $[-e, e]$. Thus order-unit normed spaces form a special class of approximate order-unit normed spaces.

(9.2) PROPOSITION. *Let $(E, C, \|.\|)$ be an ordered normed space. Then the following statements are equivalent:*

(a) *E is an order-unit normed space;*

(b) *C is 1-normal in $(E, \|.\|)$ and there exists e in E with $\|e\| \leqslant 1$ such that $e \geqslant u$ for all $u \in U$;*

(c) *U is order-convex and all elements of U are dominated by some e in E with $\|e\| \leqslant 1$;*

(d) *U is absolute-order-convex and all elements of U are dominated by some e in E with* $\|e\| \leqslant 1$;

(e) *U is solid and all elements of U are dominated by some e in E with* $\|e\| \leqslant 1$;

(f) $\|.\|$ *is a Riesz norm on* (E, C) *and all elements of U are dominated by some e in E with* $\|e\| \leqslant 1$.

Proof. (a) $\Rightarrow$ (b): If $\|.\|$ is an order-unit norm on E with order-unit e, then e dominates all elements of the open unit ball in E; also C is 1-normal by proposition (9.1). As in proposition (9.1), it is easy to verify that (b) $\Leftrightarrow$ (c) $\Rightarrow$ (d) and (e) $\Rightarrow$ (f). Next we show that (d) $\Rightarrow$ (e). Let $u \in U$. Then there exists α such that $\|u\| < \alpha < 1$. By (d), $\pm u/\alpha \leqslant e$ for some $e \in E$ with $\|e\| \leqslant 1$. Then $\alpha e \in U$ and dominates u absolutely. This shows that U is absolutely dominated. Thus the implication (d) $\Rightarrow$ (e) is clear. The implication (e) $\Rightarrow$ (d) is trivial. Therefore (d) $\Leftrightarrow$ (e). To complete the proof it remains to show that (f) $\Rightarrow$ (a). If (f) holds, then the order-interval $[-e, e]$ contains U and is contained in the closed unit ball in $(E, \|.\|)$, hence $\|.\|$ is the gauge of $[-e, e]$, and thus an order-unit norm with order-unit e.

Example. The sequence spaces ℓ^∞, c, function space L^∞, and the space of all bounded real-valued continuous functions defined on a topological space are examples of order-unit normed spaces. The space c_0 of all null sequences is an approximate order-unit normed space with no order-unit.

The concept dual to (approximate) order-unit normed spaces is that of base normed spaces. (See Edwards (1964) and Ellis (1964).) A positive subset B of an ordered vector space (E, C) is called a *base* of C if B is convex and has the property that every non-zero element c in C has a *unique* representation of the form αb with $\alpha > 0$ and $b \in B$. If, for each c in C, $h(c)$ denotes the uniquely determined positive number such that $c = h(c)b$ where $b \in B$, then h is obviously a strictly positive, additive, and positively homogeneous functional on C (h is said to be *strictly positive* if $h(c) > 0$ for all non-zero positive elements c). Let $F = C - C$. For each $x = c_1 - c_2$ in F, where $c_1, c_2 \in C$, we define

$$\tilde{h}(x) = h(c_1) - h(c_2).$$

Then $\tilde{h}$ is a well-defined linear functional on E such that $\tilde{h} \mid C = h$. Notice also that $B = \{c \in C : \tilde{h}(c) = 1\}$. By the Hahn–Banach theorem $\tilde{h}$ can be extended to define on the whole of E. Let f be a linear extension of $\tilde{h}$, then f is a strictly positive linear functional on E such that $B = C \cap f^{-1}(1) = \{c \in C : f(c) = 1\}$. This remark makes the following result clear.

(9.3) LEMMA.   *Let B be a positive set in an ordered vector space (E, C). Then B is a base of C if and only if there exists a strictly positive linear functional f on E such that $B = C \cap f^{-1}(1)$.*

If B is a base of C and if $C - C = E$, then the circled convex hull $\Gamma(B)$ of B is absorbing in E. The gauge of $\Gamma(B)$ is called a *base semi-norm* on E (defined by the base B). If the semi-norm is in fact a norm it will be referred to as a *base norm*. An ordered normed space $(E, C, \|.\|)$ is called a *base normed space* if $E = C - C$ and if there exists a base B of C such that the given norm $\|.\|$ is the gauge of $\Gamma(B)$. If B is a base of C, $E = C - C$ and if $\Gamma(B) = \mathrm{co}(B \cup -B)$ is the circled convex hull of B, then $\Gamma(B)$ is the smallest solid set containing B. In fact, it is easily seen that $\Gamma(B)$ is absolutely dominated; to see that $\Gamma(B)$ is absolute-order-convex let

$$\pm x \leqslant y = \lambda_1 b_1 - \lambda_2 b_2 \in \Gamma(B),$$

where $b_1, b_2 \in B$ and $\lambda_1, \lambda_2 \geqslant 0$ with $\lambda_1 + \lambda_2 = 1$. Since $y \pm x \geqslant 0$, there exist $b, b' \in B$ and $\alpha, \alpha' \geqslant 0$ such that $y + x = \alpha b$ and $y - x = \alpha' b'$. Then $2x = \alpha b - \alpha' b'$. To show that $x \in \Gamma(B)$, we have to show that

$$\tfrac{1}{2}(\alpha + \alpha') \leqslant 1.$$

By lemma (9.3), there exists a strictly positive linear functional f on E such that $B = C \cap f^{-1}(1)$. Notice that

$$2(\lambda_1 b_1 - \lambda_2 b_2) = 2y = \alpha b + \alpha' b',$$

hence, on applying f on both sides, we have

$$1 \geqslant \lambda_1 - \lambda_2 = \tfrac{1}{2}(\alpha + \alpha'),$$

as required. This shows that $x \in \Gamma(B)$, and hence $\Gamma(B)$ is absolute-order-convex and consequently $\Gamma(B)$ is solid. Finally, if S is a solid set containing B and if $z = \mu_1 z_1 - \mu_2 z_2 \in \Gamma(B)$, where $z_1, z_2 \in B$ and $\mu_1, \mu_2 \geqslant 0$ with $\mu_1 + \mu_2 = 1$, then let $z' = \mu_1 z_1 + \mu_2 z_2 \in B \subseteq S$. Since $\pm z \leqslant z'$ it follows that $z \in S$, showing that $\Gamma(B) \subseteq S$. Therefore we have shown that $\Gamma(B)$ is the smallest solid set in E containing B. Consequently the gauge of $\Gamma(B)$ is always a Riesz semi-norm on E.

(9.4) PROPOSITION. *Let $(E, C, \|.\|)$ be a base-normed space with base B. Then $\|.\|$ is a Riesz norm on E and the following assertions hold:*
  (a) $B = \{x \in C : \|x\| = 1\}$;
  (b) $\|.\|$ *is additive on C, i.e. $\|x+y\| = \|x\| + \|y\|$ for all x, y in C;*
  (c) *C is $(1+\varepsilon)$-generating for each $\varepsilon > 0$;*
  (d) *C is 2-normal.*

(*Remark.* Clearly the additivity of $\|.\|$ on C implies that $\|.\|$ is monotone; thus the condition is a normality condition. Similarly that U be directed upwards is a α-generating condition for appropriate α.)

Proof. That $\|.\|$ is a Riesz norm has been noted before. To show (a), let $b \in B$. Then $\|b\| \leqslant 1$. Further, suppose α is a positive real number such that $b \in \alpha\Gamma(B)$. Then $b/\alpha = \alpha'b' - \alpha''b''$ for some b', b'' in B and $\alpha', \alpha'' \geqslant 0$ with $\alpha' + \alpha'' = 1$. Let f be a strictly positive linear functional on E such that $B = C \cap f^{-1}(1)$. Then

$$\alpha' = f(\alpha'b') = f\left(\frac{b}{\alpha} + \alpha''b''\right) = \frac{1}{\alpha} + \alpha'',$$

hence $1/\alpha = \alpha' - \alpha'' \leqslant \alpha' + \alpha'' = 1$, so $\alpha \geqslant 1$. This implies that $\|b\| \geqslant 1$; consequently $\|b\| = 1$ is valid for all $b \in B$. On the other hand, if $c \in C$ and $\|c\| = 1$, then $c = \mu b^*$ for some $\mu > 0$ and $b^* \in B$; further

$$1 = \|c\| = \|\mu b^*\| = \mu\|b^*\| = \mu$$

by what we have previously proved. Therefore $c = \mu b^* = b^* \in B$. Thus (a) is verified. To verify (b), let $x, y \in C$. We can further suppose that $x \neq 0$, $y \neq 0$. Then, by (a), $x/\|x\|, y/\|y\| \in B$. Since B is convex it follows that

$$\frac{x+y}{\|x\| + \|y\|} = \frac{\|x\|}{\|x\| + \|y\|}\left(\frac{x}{\|x\|}\right) + \frac{\|y\|}{\|x\| + \|y\|}\left(\frac{y}{\|y\|}\right) \in B.$$

Applying (a) again, we have $\left\|\dfrac{x+y}{\|x\| + \|y\|}\right\| = 1$, i.e. $\|x+y\| = \|x\| + \|y\|$ as required in (b).

Let $\varepsilon > 0$ and let $x \in E$. Since $\|.\|$ is a Riesz norm, there exists $y \in E$ with $\|y\| \leqslant (1+\varepsilon)\|x\|$ such that $\pm x \leqslant y$. Let $x_1 = \frac{1}{2}(y+x)$ and $x_2 = \frac{1}{2}(y-x)$. Then $x_1, x_2 \in C$, $x = x_1 - x_2$, and

$$\|x_1\| + \|x_2\| = \|x_1 + x_2\| = \|y\| \leqslant (1+\varepsilon)\|x\|.$$

Thus (c) is proved.

Finally, we verify (d). Suppose $x \leqslant y \leqslant z$ in E. Then there exist $b_1, b_2 \in B$ and $\lambda_1, \lambda_2 \geqslant 0$ such that $y - x = \lambda_1 b_1$ and $z - y = \lambda_2 b_2$. By (b), we have

$$\|x\| + \|z\| \geqslant \|z - x\| = \|\lambda_1 b_1 + \lambda_2 b_2\| = \|\lambda_1 b_1\| + \|\lambda_2 b_2\|$$
$$= \|y - x\| + \|z - y\| \geqslant (\|y\| - \|x\|) + (\|y\| - \|z\|),$$

it follows that $\|x\| + \|z\| \geqslant \|y\|$. Consequently C must be 2-normal.

(9.5) PROPOSITION.   *Let $(E, C, \|.\|)$ be an ordered normed space. Then the following statements are equivalent:*

(a) *E is a base normed space;*

(b) *$\|.\|$ is additive on C and C is $(1+\varepsilon)$-generating for each $\varepsilon > 0$;*

(c) *$\|.\|$ is additive on C and the open unit ball U in $(E, \|.\|)$ is decomposable;*

(d) *$\|.\|$ is additive on C and U is solid;*

(e) *$\|.\|$ is a Riesz norm on (E, C) and $\|.\|$ is additive on C.*

Proof.   In view of proposition (1.7) and theorem (6.3), it is clear that (b) $\Leftrightarrow$ (c) and (d) $\Leftrightarrow$ (e). That (a) $\Rightarrow$ (b) and (a) $\Rightarrow$ (e) were proved in the preceding proposition. Therefore to complete the proof we have only to verify that (b) $\Rightarrow$ (a) and (e) $\Rightarrow$ (b).

(b) $\Rightarrow$ (a): Let $B = \{x \in C : \|x\| = 1\}$. Since $\|.\|$ is additive on C, B is convex. It is then easily verified that B is a base of C. Let $\Gamma(B) = \mathrm{co}(B \cup -B)$. Then $\Gamma(B)$ is contained in the closed unit ball of $(E, \|.\|)$. On the other hand, it contains the open unit ball U; in fact, if $\|x\| < 1$ then there exist $x_1, x_2 \in C$ with $\|x_1\| + \|x_2\| < 1$ such that $x = x_1 - x_2$. Assuming that $x_1 \neq 0$, $x_2 \neq 0$, we have $\dfrac{x_1}{\|x_1\|}, \dfrac{x_2}{\|x_2\|} \in B$ and

$$x = \|x_1\| \left(\frac{x_1}{\|x_1\|} \right) + \|x_2\| \left(\frac{x_2}{\|x_2\|} \right) \in \Gamma(B).$$

This shows that $U \subseteq \Gamma(B)$. Consequently the given norm $\|.\|$ is precisely the gauge of $\Gamma(B)$, and $(E, C, \|.\|)$ is a base normed space.

(e) $\Rightarrow$ (b): Let $x \in E$ and $\varepsilon > 0$. Since $\|.\|$ is a Riesz norm, there exists y in E with $\|y\| \leqslant (1 + \varepsilon) \|x\|$ such that $\pm x \leqslant y$. Let $x_1 = \frac{1}{2}(y + x)$ and $x_2 = \frac{1}{2}(y - x)$. Then $x = x_1 - x_2$, $x_1, x_2 \in C$, and

$$\|x_1\| + \|x_2\| = \|x_1 + x_2\| = \|y\| \leqslant (1 + \varepsilon) \|x\|,$$

showing that C is $(1 + \varepsilon)$-generating. Therefore (e) $\Rightarrow$ (b).

In what follows we shall show that the concepts of base norms and approximate order-unit norms are dual to each other. To this end, we shall first establish the fact that the additivity of the norm is dual to the directedness of the open unit ball. More precisely, we have the following theorem (cf. Ng (1969a)):

(9.6) THEOREM. *Let* $(E, C, \|.\|)$ *be an ordered normed space and* $(E', \|.\|)$ *the Banach dual space with the dual cone* C'. *Then the following statements are equivalent:*
  (a) $\|.\|$ *is additive on* C;
  (b) *the open unit ball* U' *in* E' *is directed upwards;*
  (c) *the closed unit ball* Σ' *in* E' *is directed upwards.*

Proof. (a) $\Rightarrow$ (b): Suppose that f_1, f_2 are elements of U'. Let δ be such that $\max\{\|f_1\|, \|f_2\|\} < \delta < 1$. We define

$$p(x) = \delta \|x\| \qquad (x \in E)$$
and
$$q(x) = \sup\{f_1(x_1) + f_2(x_2) : x_1, x_2 \in C, x_1 + x_2 = x\} \qquad (x \in C).$$

Then p is sublinear and q is superlinear on where they are defined. Also, since $\|.\|$ is additive, it is easily seen that $q(x) \leqslant p(x)$ for each x in C. By Bonsall's generalization of the Hahn–Banach theorem (1.15), there exists a linear functional f on E such that $q(x) \leqslant f(x)$ and $f(y) \leqslant p(y)$ for all $x \in C$ and all $y \in E$. Then $f_1, f_2 \leqslant f$ and $\|f\| \leqslant \delta < 1$ (so $f \in U'$).

(b) $\Rightarrow$ (c): Let f, g be an arbitrary pair of elements in Σ'. Then, by (b), for each positive integer n, there exists h_n in U' such that

$$f, g \leqslant \frac{n+1}{n} h_n.$$

By the Alaoglu theorem, Σ' is $\sigma(E'E)$-compact, so $\{h_n\}$ has a $\sigma(E'E)$-cluster point h_0, say. Then $h_0 \in \Sigma'$ and $f, g \leqslant h_0$. This shows that Σ' is directed upwards.

(c) $\Rightarrow$ (a): Let x, y be in C. By the Hahn–Banach theorem, there exist $f, g \in \Sigma'$ such that $f(x) = \|x\|$ and $g(y) = \|y\|$. For this pair of functions f, g, there exists $h \in \Sigma'$ such that $f, g \leqslant h$ by (c). Then

$$\|x\| + \|y\| = f(x) + g(y) \leqslant h(x) + h(y) \leqslant \|x + y\|,$$

consequently, $\|x\| + \|y\| = \|x + y\|$ by virtue of the triangle inequality for $\|.\|$.

The following result is dual to the preceding theorem and is parallel to theorems (5.16) and (6.12):

(9.7) THEOREM.   *Let C be a cone in a normed space $(E, \|.\|)$ and let $(E', \|.\|)$ be the Banach dual space with the dual cone C'. We consider the following statements:*

(a) $\|.\|$ *is additive on C';*

(b) *the open unit ball U in $(E, \|.\|)$ is directed upwards.*

Then (b) $\Rightarrow$ (a). *If, in addition, C is assumed to be $\|.\|$-complete, then* (a) $\Rightarrow$ (b), *hence* (a) *and* (b) *are equivalent. Furthermore, if C is $\|.\|$-complete and if* (a) *or* (b) *holds, then $(E, \|.\|)$ must be complete.*

(*Remark.*   It is easy to see that if the closed unit ball is directed upwards then so is the open unit ball. The converse is incorrect: for a counterexample see Asimow (1968). In the case where X and C are assumed to be $\|.\|$-complete, the theorem was proved independently by Asimow (1968) and Ng (1969a). A more general theorem than the present form was announced by Ng and Duhoux (1973).)

Proof.   If (a) or (b) holds, then we established the fact in theorem (5.16), that $(E, \|.\|)$ is complete whenever C is $\|.\|$-complete.

(b) $\Rightarrow$ (a): Let $f, g \in C'$ and let $x, y \in U$. By (b), there exists $z \in U$ such that $x, y \leqslant z$. Then

$$f(x) + g(y) \leqslant f(z) + g(z) \leqslant \|f + g\| \|z\| \leqslant \|f + g\|.$$

Since x, y are arbitrary in U, it follows that $\|f\| + \|g\| \leqslant \|f + g\|$; consequently $\|f\| + \|g\| = \|f + g\|$ by virtue of the triangle inequality.

Finally we show that (a) $\Rightarrow$ (b) under the assumption that C is $\|.\|$-complete (hence E is also $\|.\|$-complete). Let x', x'' be in U. Let δ be such that $\max\{\|x'\|, \|x''\|\} < \delta < 1$. We shall find an $x \in U$ such that $x \geqslant x', x''$. To do this, let

$$P(f) = \delta \|f\| \qquad (f \in E')$$

and

$$Q(f) = \sup\{f'(x') + f''(x'') : f', f'' \in C', f = f' + f''\} \qquad (f \in C').$$

Then P is sublinear and Q is superlinear in the domains where they are defined; also, by (a), $Q(f) \leqslant P(f)$ for each f in C'. Further, P is lower $\sigma(E', E)$-semi-continuous on E. Next we show that Q is upper semi-continuous on $\Sigma' \cap C'$ under the relative $\sigma(E', E)$-topology. If not, then there exist a real number λ and a net $\{f_\alpha\}$ in $\Sigma' \cap C'$ convergent to f_0 such that $Q(f_\alpha) > \lambda > Q(f_0)$ for each α. Then, for each α, there exist f'_α, f'' in C' with $f_\alpha = f'_\alpha + f''_\alpha$ such that

$$f'_\alpha(x') + f''_\alpha(x'') > \lambda.$$

Notice that $0 \leqslant f'_\alpha \leqslant f_\alpha$, so $\|f'_\alpha\| \leqslant \|f_\alpha\| \leqslant 1$ by (a). This shows that each f'_α is in $\Sigma' \cap C'$. Similarly $f''_\alpha \in \Sigma' \cap C'$. Since $\Sigma' \cap C'$ is $\sigma(E', E)$-compact, $\{f'_\alpha\}$, $\{f''_\alpha\}$ have convergent subnets with limits, say f'_0 and f''_0, in $\Sigma' \cap C'$ respectively. Since $f_\alpha = f'_\alpha + f''_\alpha$, we must have $f_0 = f'_0 + f''_0$. Passing to the limit in the last displayed inequality, we have

$$Q(f_0) \geqslant f'_0(x') + f''_0(x'') = \lim(f'_\alpha(x') + f''_\alpha(x'')) \geqslant \lambda,$$

contrary to the fact that $\lambda > Q(f_0)$. The contradiction shows that Q is upper semi-continuous on $\Sigma' \cap C'$. By the separation theorem given in corollary (2.19), for $0 < \varepsilon < 1 - \delta$, there exists x in E such that

$$Q(f) \leqslant f(x) \qquad (f \in C')$$
and
$$f(x) \leqslant P(f) + \varepsilon \|f\| \qquad (f \in E').$$

By the last inequality, it follows from the Hahn–Banach theorem that

$$\|x\| = \sup\{f(x) : f \in E', \|f\| \leqslant 1\} \leqslant \delta + \varepsilon < 1.$$

Also, for each f in C', $f(x') \leqslant Q(f) \leqslant f(x)$. Since C is complete (hence closed), it follows from the Hahn–Banach separation theorem (cf. theorem (2.15)) that $x' \leqslant x$. Similarly $x'' \leqslant x$. Thus x is an element of U which dominates x', x''.

The following theorem is due to Krein and Ellis (cf. Ellis (1964)).

(9.8) THEOREM. *Let $(E, C, \|.\|)$ be an ordered normed space and let $(E', \|.\|)$ be the Banach dual space with the dual cone C'. We consider the following statements:*
 (a) *$(E, C, \|.\|)$ is a base normed space;*
 (b) *$(E', C', \|.\|)$ is an approximate order-unit normed space;*
 (c) *$(E', C', \|.\|)$ is an order-unit normed space.*
Then (a) $\Rightarrow$ (b) $\Leftrightarrow$ (c). If C is assumed to be $\|.\|$ complete, then (c) $\Rightarrow$ (a) and hence (a), (b), and (c) are mutually equivalent. Further, if C is $\|.\|$-complete and if one of the statements (a), (b), and (c) holds, then $(E, \|.\|)$ must be complete.

Proof. (a) $\Rightarrow$ (b): If $(E, C, \|.\|)$ is a base normed space, then $\|.\|$ is a Riesz norm on (E, C) and is additive on C by proposition (9.4). It follows from theorem (6.12) and theorem (9.6) that the norm on the Banach dual space (E', C') is also a Riesz norm and the open unit ball U' is directed upwards; hence $(E', C', \|.\|)$ is an approximate order-unit normed space by proposition (9.1).

(b) $\Rightarrow$ (c): If (b) holds, then the open unit ball U' in $(E', \|.\|)$ is directed upwards. For each $u \in U'$, let $e_u = u$. Then $\{e_u : u \in U'\}$ is an increasing net in the closed unit ball Σ' which is $\sigma(E', E)$-compact by the Alaoglu theorem. Let e be a cluster point of $\{e_u\}$. Then $e \geqslant e_u$ for all u. In view of proposition (9.2), we conclude that $(E', C', \|.\|)$ is an order-unit normed space.

Finally we show that if C is $\|.\|$-complete and if (c) holds then $(E, \|.\|)$ is complete and (a) holds. By (c), C' is 1-normal in $(E', \|.\|)$. Hence, since C is $\|.\|$-complete, it follows from theorem (5.16) that $(E, \|.\|)$ must be complete and C is $(1+\varepsilon)$-generating for each $\varepsilon > 0$. Furthermore, the open unit ball in the order-unit normed space $(E', C', \|.\|)$ is directed upwards, and it follows from theorem (9.6) that the norm $\|.\|$ on E is additive on C. This, together with the $(1+\varepsilon)$-generating property of C, implies that $(E, C, \|.\|)$ is a base normed space by proposition (9.5), i.e. (a) holds.

The following theorem is dual to theorem (9.8) and is due to Ng (1969).

(9.9) THEOREM. *Let $(E, C, \|.\|)$ be an ordered normed space and let $(E', C', \|.\|)$ be the Banach dual space with the dual cone C'. We consider the following statements:*

(a) *$(E, C, \|.\|)$ is an approximate order-unit normed space;*

(b) *$(E', C', \|.\|)$ is a base normed space.*

Then (a) $\Rightarrow$ (b). If C is assumed to be $\|.\|$-complete then (b) $\Rightarrow$ (a) and hence (a) and (b) are equivalent. Further, if C is $\|.\|$-complete and if either (a) or (b) holds then $(E, \|.\|)$ must be complete.

Proof. If C is $\|.\|$-complete, then we have the following chain of equivalences (cf. proposition (9.1), proposition (9.5), theorem (6.12), and theorem (9.7)):

(a) $\Leftrightarrow$ the open unit ball is directed upwards and $\|.\|$ is a Riesz norm on E

$\Leftrightarrow$ the norm on E' is additive on C' and is a Riesz norm on $E' \Leftrightarrow$ (b). The other assertions can also be easily verified by virtue of the propositions quoted.

In general, that E' be a base normed space does not imply that E is an order-unit normed space, as the following example shows.

Example. Let c_0 be the Banach lattice of all null sequences of real numbers (with the natural ordering and the supremum-norm). Then the Banach dual space of c_0 is the space ℓ^1 of all summable sequences of real numbers. The norm on c_0 is a Riesz norm and the open unit ball is directed upwards; so c_0 is an approximate order-unit normed space and ℓ^1 is a base normed space. Further c_0 has no order-unit. In fact, consider an element $x = (x_1, x_2, ...)$ of c_0. Let $y = (\sqrt{x_1}, \sqrt{x_2}, ...)$. Then $y \in c_0$, and $\lim M\sqrt{x_n} = 0$ for any constant $M > 0$. Hence there does not exist a constant M with the property that

$$(\sqrt{x_1}, \sqrt{x_2}...) = y \leqslant Mx = (Mx_1, Mx_2, ...).$$

This shows that the order-interval $[-x, x]$ does not absorb y, so x is *not* an order-unit. Thus c_0 has no order-unit at all.

In the situation of the preceding theorem, it is easily seen that the set $B = \{f \in C' : \|f\| = 1\}$ is a base of C' and the defining base norm is the same as the given norm. It turns out that the existence of an order-unit in E is related to a topological property of B.

(9.10) THEOREM (Edwards). *Let $(E, C, \|.\|)$ be an ordered normed space, and let $(E', C', \|.\|)$ be the Banach dual space with dual cone C'. We consider the following statements:*

(a) $(E, C, \|.\|)$ *is an order-unit normed space;*

(b) $(E', C', \|.\|)$ *is a base normed space with a $\sigma(E', E)$-compact base B defining the norm $\|.\|$.*

Then (a) $\Rightarrow$ (b). *If C is assumed to be $\|.\|$-complete, then* (b) $\Rightarrow$ (a); *hence the statements* (a) *and* (b) *are equivalent. Moreover, if C is $\|.\|$-complete and if either* (a) *or* (b) *holds, then $(E, \|.\|)$ must be complete.*

Proof. The final assertion can be proved easily, as before. Next we show that (a) $\Rightarrow$ (b). Suppose $(E, C, \|.\|)$ has an order-unit e defining the order-unit norm $\|.\|$ on E. By the preceding theorem, $(E', C', \|.\|)$ is a base normed space; hence the set

$$B = \{f \in C' : \|f\| = 1\}$$

is a base of C' and the corresponding base norm is precisely the norm $\|.\|$ on E'. It is not difficult to see that

$$B = \{f \in C' : \|f\| \leqslant 1 \quad \text{and} \quad f(e) = 1\}.$$

By the Alaoglu theorem, B is $\sigma(E', E)$-compact, proving (b).

Conversely, suppose (b) holds and that C is $\|.\|$-complete. Then, by theorem (9.9), $(E, \|.\|)$ is also complete and is an approximate order-unit normed space. By (b), co$(B \cup -B)$ contains the open unit ball U' and is contained in the closed unit ball Σ' in $(E', \|.\|)$; further, since B is $\sigma(E', E)$-compact, it is easily seen that co$(B \cup -B)$ must be equal to Σ'. Next, we note that each $f \in E'$ can be expressed as $f = f_1 - f_2$ for some $f_1, f_2 \in C'$; define

$$\mu(f) = \|f_1\| - \|f_2\|.$$

Since $\|.\|$ is additive on C', $\mu(f)$ does not depend on the particular choice of the components f_1, f_2 of f. Hence $f \to \mu(f)$ is a well-defined linear functional on E', and its restriction to C' is identical to that of $\|.\|$. Notice that

$$\tfrac{1}{2}B - \tfrac{1}{2}B = \Sigma' \cap \mu^{-1}(0). \tag{9.1}$$

In fact, the set on the left-hand side is certainly contained in that on the right-hand side; on the other hand, let $f \in \Sigma' \cap \mu^{-1}(0)$ and write $f = \lambda_1 b_1 - \lambda_2 b_2$, where $b_1, b_2 \in B$ and $\lambda_1, \lambda_2 \geqslant 0$ with $\lambda_1 + \lambda_2 = 1$. Then

$$0 = \mu(f) = \|\lambda_1 b_1\| - \|\lambda_2 b_2\| = \lambda_1 - \lambda_2,$$

it follows that $\lambda_1 = \tfrac{1}{2} = \lambda_2$. Therefore $f = \tfrac{1}{2}f_1 - \tfrac{1}{2}f_2 \in \tfrac{1}{2}B - \tfrac{1}{2}B$. Thus formula (9.1) is proved. Since B is $\sigma(E', E)$-compact so is the set $\tfrac{1}{2}B - \tfrac{1}{2}B$. Therefore the kernel $\mu^{-1}(0)$ of μ intersects Σ' in a $\sigma(E', E)$-compact set. By the Krein–Smulian theorem (cf. e.g. Schaefer (1966, p. 152)), $\mu^{-1}(0)$ is $\sigma(E', E)$-closed and μ is $\sigma(E', E)$-continuous. Consequently, there exists e in E such that $\mu(f) = f(e)$ for all $f \in E'$. As noted before, $(E, C, \|.\|)$ is, at least, an approximate order-unit normed space: thus, in view of proposition (9.2), to complete the proof we have only to show that $\|e\| \leqslant 1$ and $e \geqslant x$ whenever $\|x\| < 1$. To verify this, let $f = \lambda_1 b_1 - \lambda_2 b_2 \in \Sigma' = \mathrm{co}(B \cup -B)$ for appropriate b_1, b_2 and λ_1, λ_2. Then

$$f(e) = \mu(f) = \|\lambda_1 b_1\| - \|\lambda_2 b_2\| = \lambda_1 - \lambda_2 \leqslant 1.$$

By the Hahn–Banach theorem, it follows that $\|e\| \leqslant 1$. Moreover if $x \in E$ and $\|x\| < 1$ then

$$f(x) \leqslant 1 = \|f\| = \mu(f) = f(e)$$

for all $f \in B$; and so $f(x) \leqslant f(e)$ for all $f \in C'$. Since C is $\|.\|$-complete (hence closed), it follows that $x \leqslant e$, whenever $\|x\| < 1$. This completes the proof of the theorem.

Let K be a compact convex subset of a locally convex space. A real-valued function a on K is said to be *affine* if

$$a(\lambda_1 k_1 + \lambda_2 k_2) = \lambda_1 a(k_1) + \lambda_2 a(k_2)$$

whenever $k_1, k_2 \in K$ and $\lambda_1, \lambda_2 \geqslant 0$ with $\lambda_1 + \lambda_2 = 1$. Let $C(K)$ denote the ordered Banach space of all real-valued continuous functions on K, and let $A(K)$ be the closed subspace of $C(K)$ consisting of all affine functions. Then the constant function 1 is an order-unit in $C(K)$ and defines the order-unit norm which is precisely the usual supremum in

$C(K)$. Further, since $1 \in A(K)$, $A(K)$ is also an order-unit normed space in its own right. Thus the implication (b) $\Rightarrow$ (a) in the following theorem is clear.

(9.11) THEOREM. *Let $(E, C, \|.\|)$ be an ordered Banach space with a closed cone C. Then the following statements are equivalent:*

(a) *$(E, C, \|.\|)$ is an order-unit normed space;*

(b) *there exists a compact convex set K in a locally convex space such that $(E, C, \|.\|)$ is isometrically order-isomorphic to $A(K)$.*

Proof. It remains to show (a) $\Rightarrow$ (b). Let

$$B = \{f \in E' : \|f\| = 1, f \in C'\}.$$

In the proof of the preceding theorem, we noted that B is a $\sigma(E', E)$-compact base of C', and $\Gamma(B) = \Sigma'$. Let $A(B)$ be the space of all affine $\sigma(E', E)$-continuous real-valued functions on B. We shall represent $(E, C, \|.\|)$ as $A(B)$.

For each x in E, define $\hat{x} \in A(B)$ by the rule

$$\hat{x}(f) = f(x) \qquad (f \in B).$$

Since $\Gamma(B) = \Sigma'$, it follows from the Hahn–Banach theorem that $\|x\| = \|\hat{x}\|$. Further, since C is closed, it is easily seen that $x \geqslant 0$ if and only if $\hat{x} \geqslant 0$. It is now clear that the map $x \to \hat{x}$ is isometrically order-isomorphic from E into $A(B)$. To complete the proof we have to show that the map is onto. Let $a \in A(B)$. Since B is a base of C', a can be uniquely extended to become an affine function on C', and consequently a can be uniquely extended to become a linear functional on $C' - C' = E'$. Notice that if $f \in \Sigma' = \Gamma(B)$ and $f = \lambda b - \mu c$, where $b, c \in B$ and $\mu, \lambda \geqslant 0$ with $\lambda + \mu = 1$, then

$$a(f) = \lambda a(b) - \mu a(c).$$

Since $\Sigma' = \Gamma(B)$ and a is continuous on the $\sigma(E', E)$-compact set B, it follows easily that a is continuous on Σ'. In particular, the intersection $a^{-1}(0) \cap \Sigma'$ of the kernel with the closed unit ball is $\sigma(E', E)$-closed; consequently, it follows from the Krein–Smulian theorem that $a^{-1}(0)$ is $\sigma(E', E)$-closed, and hence a is a $\sigma(E', E)$-continuous linear functional on E'. Therefore there exists x in E such that $\hat{x} = a$; this shows that the map $x \to \hat{x}$ is onto $A(B)$, and completes the proof of the theorem.

In order to represent approximate order-unit normed spaces in the spirit of the preceding theorem we need to study a special type of compact convex sets. Let $(E, C, \mathscr{P})$ be an ordered convex space. A non-empty $\mathscr{P}$-compact convex subset K of C is called a *cap* of C if $C \backslash K$ is also convex. Notice that any cap necessarily contains the origin. K is said to be *universal* if $C = \text{pos } K$; i.e. if $C = \bigcup \{\lambda K : \lambda > 0\}$. If C has a compact base B, for instance, then the set $\{\lambda b : 0 \leqslant \lambda \leqslant 1, b \in B\}$ is a universal cap of C. Another example of a universal cap is the positive part of the unit ball in the Banach dual space of an approximate order-unit normed space, as the following lemma shows.

(9.12) LEMMA. *Let $(E, C, \|.\|)$ be an approximate order-unit normed space, and let $(E', C', \|.\|)$ be the Banach dual space with the dual cone C' and with the closed unit ball Σ'. Let $K = \Sigma' \cap C'$. Then K is a universal cap of C' in the locally convex space $(E', \sigma(E', E))$.*

Proof. By the Alaoglu theorem, K is a compact convex subset of E' with respect to the $\sigma(E', E)$-topology. Also, by theorem (9.9), $(E', C', \|.\|)$ is a base normed space, so $\|.\|$ is additive on C'. Consequently $C' \backslash K$ is convex. Therefore K is a cap of C'. Finally it is easily seen that K must be universal in C'.

If K is a cap, $A_0(K)$ denotes the subspace of $A(K)$ consisting of all functions vanishing at the origin 0. Then $A_0(K)$ is an ordered Banach space in its own right.

(9.13) PROPOSITION. *Let $(E, C, \|.\|)$ be an approximate order-unit normed space, and suppose that $(E, \|.\|)$ and C are complete. Then there exists a universal cap K such that $(E, C, \|.\|)$ is isometrically order-isomorphic to $A_0(K)$, where K may be taken to be the positive part of the unit ball in the Banach dual space E'.*

Proof. Let $K = \Sigma' \cap C'$, as in the preceding lemma. Then K is a universal cap of C' in $(E', \sigma(E', E))$. For each x in E define $\hat{x} \in A_0(K)$ by

$$\hat{x}(f) = f(x) \qquad (f \in K).$$

Then, as in the proof of theorem (9.11), $x \to \hat{x}$ is an isometrically order-isomorphic map of E onto $A_0(K)$.

In what follows we shall show that the converse of proposition (9.13) holds. To this end, we first prove the converse of lemma (9.12).

(9.14) PROPOSITION. *Let $(E, C, \mathscr{P})$ be an ordered convex space such that C is $\mathscr{P}$-closed and $E = C - C$. Let K be a universal cap of C, and let $\|.\|$ be the gauge of $\mathrm{co}(K \cup -K)$. Then $(E, C, \|.\|)$ is a base normed space, and there exists an approximate order-unit normed space $(V, W, \|.\|)$ such that $(E, C, \|.\|)$ is isometrically order-isomorphic to the ordered Banach dual space $(V', W', \|.\|)$ of $(V, W, \|.\|)$. Furthermore, if $\Sigma = \{x \in E : \|x\| \leqslant 1\}$ then*

$$K = \Sigma \cap C. \tag{9.2}$$

Proof. For each $x \in C$, we define

$$p(x) = \inf\{\lambda > 0 : x \in \lambda K\}.$$

Since K is $\mathscr{P}$-compact convex and since $C \backslash K$ is convex, it is not difficult to verify that p is a positively homogeneous, additive functional on C and $K = \{x \in C : p(x) \leqslant 1\}$. Moreover, since K is $\mathscr{P}$-compact, $p(x) \neq 0$ whenever $x \in C$ and $x \neq 0$. Let $B = \{x \in C : p(x) = 1\}$. Then it is easily seen that B is a base of C and $\mathrm{co}(K \cup -K) = \mathrm{co}(B \cup -B)$. Hence the gauge $\|.\|$ of $\mathrm{co}(K \cup -K)$ is precisely the base semi-norm defined by the base B. Furthermore, since K is $\mathscr{P}$-compact, $\mathrm{co}(K \cup -K)$ is $\mathscr{P}$-compact, and hence $\|.\|$ is in fact a norm and

$$\mathrm{co}(B \cup -B) = \mathrm{co}(K \cup -K) = \{x \in E : \|x\| \leqslant 1\},$$

that is, $\mathrm{co}(B \cup -B) = \Sigma$. To verify formula (9.2), let $x \in \Sigma \cap C$, and suppose that $x = \lambda_1 b_1 - \lambda_2 b_2$, where $b_1, b_2 \in B$ and $\lambda_1, \lambda_2 \geqslant 0$ with $\lambda_1 + \lambda_2 = 1$. Then

$$p(x) + \lambda_2 = p(x + \lambda_2 b_2) = p(\lambda_1 b_1) = \lambda_1,$$

so $p(x) = \lambda_1 - \lambda_2 \leqslant 1$ and $x \in K$. This shows that $\Sigma \cap C \subseteq K$; consequently $\Sigma \cap C = K$ since it is obvious that $\Sigma \cap C \supseteq K$. Thus formula (9.2) is proved. We have shown that $(E, C, \|.\|)$ is a base normed space with the $\mathscr{P}$-compact closed unit ball Σ, and that K is the positive part $\Sigma \cap C$ of the unit ball Σ. Finally we show that $(E, \|.\|)$ may be identified with a Banach dual space. Accordingly, let $(E, \mathscr{P})'$ and $(E, \|.\|)'$ denote the dual spaces of E und $\mathscr{P}$ and $\|.\|$ respectively. Let V be the space of all linear functionals f on E such that f is $\mathscr{P}$-continuous on Σ. Then

$$(E, \mathscr{P})' \subseteq V \subseteq (E, \|.\|)'. \tag{9.3}$$

The first inequality is obvious. To see the second inequality let $f \in V$, then $f(\Sigma)$ is the continuous image of the $\mathscr{P}$-compact set Σ, so is compact and hence bounded in $\mathbf{R}$. Therefore f is continuous on $(E, \|.\|)$, and formula (9.3) is proved. Now it is easily seen that V is a closed subspace of the Banach space $(E, \|.\|)'$. Thus V may be regarded as a Banach space in its own right. Let W be the cone in V consisting of all positive linear functionals on E. Then $(V, W, \|.\|)$ is an ordered Banach space and W is $\|.\|$-closed. Let $(V', W', \|.\|)$ denote the Banach dual of $(V, \|.\|)$ with the dual cone W'. For each x in E, define $\psi(x)$ by the rule

$$(\psi(x))(v) = v(x) \qquad (v \in V).$$

Then it is easy to see that ψ is a 1–1 continuous (in fact norm-reducing) map from E into the Banach dual space V' of V. Since the cone C in E is $\mathscr{P}$-closed, it is easy to see that $x \geqslant 0$ in E if and only if $\psi(x) \geqslant 0$ in V'. Also, since each v in V is $\mathscr{P}$-continuous on Σ, the restriction $\psi|\Sigma$ of ψ to Σ is continuous with respect to the relative $\mathscr{P}$-topology in Σ and the $\sigma(V', V)$-topology in V'. Since Σ is $\mathscr{P}$-compact it follows that $\psi(\Sigma)$ is $\sigma(V', V)$-compact. Also the set $\psi(\Sigma)$ is convex. By the bipolar theorem, it is precisely its bipolar $(\psi(\Sigma))^{\pi\pi}$ with respect to the duality $\langle V', V \rangle$. Note that

$$(\psi(\Sigma))^{\pi} = \{v \in V : (\psi(x))(v) \leqslant 1 \ \forall x \in \Sigma\}$$

$$= \{v \in V : v(x) \leqslant 1 \ \forall x \in \Sigma\},$$

which is the unit ball in V, and hence $(\psi(\Sigma))^{\pi\pi}$ (that is, $\psi(\Sigma)$) is the unit ball in V'. In other words, ψ maps Σ onto the unit ball in V'. Therefore ψ is an isometric order-isomorphism from $(E, C, \|.\|)$ onto the Banach dual space $(V', W', \|.\|)$. Since $(E, C, \|.\|)$ is a base normed space, so is $(V', W', \|.\|)$ and it follows from theorem (9.9) that $(V, W, \|.\|)$ is an approximate order-unit normed space.

    Remark. The condition that C is $\mathscr{P}$-closed is in fact automatically satisfied in view of formula (9.2) and the Krein–Smulian theorem.

(9.15) THEOREM. *Let $(E, C, \|.\|)$ be an ordered Banach space with a closed cone C. Then the following statements are equivalent:*

    (a) *$(E, C, \|.\|)$ is an approximate order-unit normed space;*

    (b) *there exists a universal cap K of a cone P such that $(E, C, \|.\|)$ is isometrically order-isomorphic to $A_0(K)$.*

    Proof. The implication (a) $\Rightarrow$ (b) was established in proposition (9.13). Conversely, suppose K is a universal cap of a cone P, and let $P - P = F$. Let $\|.\|$ be the gauge of $\mathrm{co}(K \cup -K)$. Then, by the preceding proposition, there exists an approximate order-unit normed space $(V, W, \|.\|)$ such that $(F, P, \|.\|)$ may be identified with the ordered Banach dual space $(V', W', \|.\|)$, and K is identified with the positive part of the unit ball in $(V', W', \|.\|)$ (cf. formula (9.2) in proposition (9.14)). By proposition (9.13), $(V, W, \|.\|)$ is isometrically order-isomorphic to $A_0(K)$. Thus $A_0(K)$ is also an approximate order-unit normed space. This proves that (b) $\Rightarrow$ (a) and completes the proof of the theorem.

We recall that a semi-norm p on an ordered vector space (E, C) is called a Riesz semi-norm if it satisfies the following two conditions:

(i) absolute-monotonicity—if $-y \leqslant x \leqslant y$ then $p(x) \leqslant p(y)$;

(ii) for any $x \in E$ with $p(x) < 1$ there exists $y \in C$ with $p(y) < 1$ such that $-y \leqslant x \leqslant y$.

Now suppose that (E, C) is a Riesz space (i.e. vector lattice), and that p is a Riesz semi-norm. Then, in view of the absolute-monotonicity property of p and $-|x| \leqslant x \leqslant |x|$, we have $p(x) \leqslant p(|x|)$. If there is x in E such that $p(x) < p(|x|)$, and we can assume without loss of generality that $p(x) < 1 < p(|x|)$, then there exists $y \in C$ with $p(y) < 1$ such that $-y \leqslant x \leqslant y$, and thus $-y \leqslant |x| \leqslant y$; using the absolute-monotone property of p again, we obtain $p(|x|) \leqslant p(y) < 1$, contrary to the fact that $1 < p(|x|)$. This contradiction shows that $p(x) = p(|x|)$ for all $x \in E$. This remark makes the following result clear.

(9.16) LEMMA. *Let (E, C) be a Riesz space and p a semi-norm on E. Then the following statements are equivalent:*

(a) *p is a Riesz semi-norm;*

(b) *p is monotone and $p(|x|) = p(x)$ for all $x \in E$;*

(c) *if $|x| \leqslant |y|$ then we have $p(x) \leqslant p(y)$.*

Let (E, C) be a vector lattice. A Riesz semi-norm on E is called a *Riesz norm* if it is, in fact, a norm. A vector lattice equipped with a Riesz norm is called a *normed lattice* (or a *normed Riesz space*). If the norm in a normed lattice is complete then the normed lattice is called a *Banach lattice* (or briefly *B-lattice*). It is elementary and well known (cf. proposition (10.3), in the next chapter) that $|x^+ - y^+| \leqslant |x - y|$ where x, y are in E; hence, if $\|.\|$ is a Riesz norm on E, then

$$\|x^+ - y^+\| \leqslant \|x - y\|.$$

This implies that the map $x \to x^+$ is continuous in a normed lattice $(E, C, \|.\|)$. Similarly we can show that the lattice operations $x \to x^-$ and $x \to |x|$ are continuous. Consequently the positive cone in a normed

8

vector lattice must be closed (and hence the positive cone in a B-lattice must be complete). In Chapter 11 we shall generalize the theory of normed vector lattices to a more general case of so-called locally convex Riesz spaces. Nevertheless, in the remainder of this chapter we study some results peculiar to the normable case.

A Banach lattice $(X, C, \|.\|)$ is called:

(i) an *AM-space* (or AL^∞-*space*) if $\|x_1 \vee x_2\| = \|x_1\| \vee \|x_2\|$ for each pair of elements x_1, x_2 in C;

(ii) an *AL-space* (or AL^1-*space*) if $\|x_1 + x_2\| = \|x_1\| + \|x_2\|$ for each pair of elements x_1, x_2 in C;

(iii) an AL^p-*space* $(1 < p < \infty)$ if

$$\|x_1 \vee x_2\|^p \leqslant \|x_1\|^p + \|x_2\|^p \leqslant \|x_1 + x_2\|^p$$

for each pair of elements x_1, x_2 in C.

The following result characterizes AM-spaces.

(9.17) PROPOSITION. *Let* $\|.\|$ *be a norm on a vector lattice* (X, C) *(Riesz space) and suppose that* $(X, \|.\|)$ *is complete. Then the following statements are equivalent:*

(a) $(X, C, \|.\|)$ *is an AM-space;*

(b) $\|.\|$ *is an approximate order-unit norm on* (X, C);

(c) $(X, C, \|.\|)$ *is a Banach lattice and the closed unit ball* Σ *in* $(X, \|.\|)$ *is directed upwards;*

(d) $(X, C, \|.\|)$ *is a Banach lattice and the open unit ball* U *in* $(X, \|.\|)$ *is directed upwards.*

Proof. It is easy to see that (a) $\Rightarrow$ (c) $\Rightarrow$ (d). In view of lemma (9.16) and proposition (9.1) it is clear that (b) $\Leftrightarrow$ (d). Thus it remains to show (d) $\Rightarrow$ (a). Let $x_1, x_2 \in C$. Then $0 \leqslant x_1, x_2 \leqslant x_1 \vee x_2$; by (d), it follows that $\|x_1\|, \|x_2\| \leqslant \|x_1 \vee x_2\|$; hence $\|x_1\| \vee \|x_2\| \leqslant \|x_1 \vee x_2\|$. We show further that the strict inequality cannot hold. Otherwise there exists a real number M such that $\|x_1\| \vee \|x_2\| < M < \|x_1 \vee x_2\|$. Since U is directed upwards, there exists $x \in X$ with $\|x\| < M$ such that $x \geqslant x_1, x_2$. Then $0 \leqslant x_1 \vee x_2 \leqslant x$ and $\|x_1 \vee x_2\| \leqslant \|x\| < M$, a contradiction. Therefore we must have $\|x_1 \vee x_2\| = \|x_1\| \vee \|x_2\|$ whenever $x_1, x_2 \in C$. This implies that X is an AM-space.

The following result characterizes AL-spaces, and is dual to the preceding proposition.

(9.18) PROPOSITION.   *Let* $\|\,.\,\|$ *be a norm on a vector lattice* (X, C) *and suppose that* $(X, \|\,.\,\|)$ *is complete. Then the following statements are equivalent:*

(a) $(X, C, \|\,.\,\|)$ *is an AL-space;*

(b) $\|\,.\,\|$ *is a Riesz norm on* (X, C) *and is additive on* C;

(c) $\|\,.\,\|$ *is a base norm on* (X, C).

Proof.   By lemma (9.16), (a) $\Leftrightarrow$ (b). By proposition (9.5), (b) $\Leftrightarrow$ (c).

Let $(E, C, \|\,.\,\|)$ be an ordered Banach space for which C is closed, and we suppose further that $\|\,.\,\|$ is a Riesz norm on E. If, in addition, (E, C) is assumed to have the Riesz decomposition property, then $E^{\mathrm{b}} = E^{\#} = E'$ and E' is a vector lattice by corollary (8.3). Consequently E' is a Banach lattice. In what follows we shall show the converse: if E' is a Banach lattice then E must have the Riesz decomposition property. In order to prove this result let us return to examine the proof of the Riesz theorem once again, where (E, C) is an ordered vector space with the Riesz decomposition. The principal construction in the proof is to define, for $f_1, f_2 \in E^{\mathrm{b}}$, that

$$(f_1 \vee f_2)(x) = \sup\{f_1(x_1) + f_2(x_2) : x = x_1 + x_2,\, x_1, x_2 \in C\} \qquad (x \in C)$$

and that

$$(f_1 \wedge f_2)(x) = \inf\{f_1(x_1) + f_2(x_2) : x = x_1 + x_2,\, x_1, x_2 \in C\} \qquad (x \in C).$$

Then, because E has the Riesz decomposition property, $f_1 \vee f_2$ and $f_1 \wedge f_2$ are positively homogeneous and additive (i.e. affine) on C. Consequently, $f_1 \vee f_2$ (or, more precisely, the linear extension of $f_1 \vee f_2$) is the supremum of f_1 and f_2, and $f_1 \wedge f_2$ is the infimum of f_1, f_2. The construction can be adopted in an ordered vector space even without the Riesz decomposition property.

Let us consider an arbitrary ordered vector space (E, C). Let p_1, p_2 be two sublinear functionals on C and suppose they are bounded on order-intervals in C. Define

$$(p_1 \wedge p_2)(x) = \inf\{p_1(x_1) + p_2(x_2) : x = x_1 + x_1,\, x_1, x_1 \in C\} \qquad (x \in C).$$

Then $p_1 \wedge p_2$ is an 'infimum' of p_1, p_2; that is, $p_1 \wedge p_2$ is the greatest sublinear and real-valued functional on C which is smaller than (or equal to) both p_1, p_2. Similarly, if q_1, q_2 are two superlinear functionals on C and bounded on each order-interval in C, then the functional

$q_1 \vee q_2$ defined by

$$(q_1 \vee q_2)(x) = \sup\{q_1(x_1) + q_2(x_2) : x = x_1 + x_2, \, x_1, \, x_2 \in C\} \qquad (x \in C)$$

is a 'supremum' of q_1, q_2; that is, $q_1 \vee q_2$ is the smallest superlinear functional on C which is larger than q_1, q_2. We remark that, even if $q_1, q_2 \leqslant p_1, p_2$, it is not necessarily true that $q_1 \vee q_2 \leqslant p_1 \wedge p_2$. Of course, this will be the case if p_i, q_i are all affine (that is, both sublinear and superlinear).

The following result sharpens corollary (2.19) in the presence of the Reisz decomposition property.

(9.19) THEOREM. *Let $(E, C, \|\cdot\|)$ be an ordered Banach space with closed cone C, and suppose that C is normal and generating in $(E, \|\cdot\|)$. Let p, q be linear functionals on the dual cone C' with the following properties:*

(a) *$q \leqslant p$ on C', i.e. $q(f) \leqslant p(f)$ for all $f \in C'$;*

(b) *q is upper semi-continuous on $C' \cap \Sigma'$ with respect to the $\sigma(E', E)$-topology, where $\Sigma' = \{f \in E' : \|f\| \leqslant 1\}$;*

(c) *p is lower semi-continuous and bounded on $C' \cap \Sigma'$.*

If the ordered Banach dual space (E', C') has the Riesz decomposition property, then there exists x in E such that

$$q(f) \leqslant f(x) \leqslant p(f), \quad \text{for all } f \in C'.$$

(*Remark.* In view of the Krein–Smulian theorem, (b) implies that q is, in fact, upper semi-continuous on the whole of C' with respect to the $\sigma(E', E)$-topology. Similarly (c) implies that p is lower semi-continuous on C'. Let $\hat{E}$ denote the canonical image in the second dual E'' of E. Then the conclusion of the theorem states that there is a $\sigma(E', E)$-continuous linear functional $\hat{x}$ lying between the semi-continuous functionals p and q.)

Proof. By corollary (3.2) and proposition (5.6), there exist positive real numbers α, β such that C is α-normal and β-generating in $(E, \|\cdot\|)$. Hence, by corollary (6.11), there exists a Riesz norm $\|\cdot\|'$ on (E, C) equivalent to the given norm $\|\cdot\|$. Therefore we can assume without loss of generality that $\|\cdot\|$ is a Riesz norm on E (otherwise, consider $\|\cdot\|'$ instead of $\|\cdot\|$); hence the norm $\|\cdot\|$ on the dual space E' is also a Riesz norm.

By corollary (2.19), there exists x_0 in E such that

$$q(f) \leqslant f(x_0) \leqslant p(f) + \frac{\|f\|}{2} \qquad (f \in C'). \qquad (9.4)$$

Let $\hat{x}_0$ denote the canonical image in E'' of x_0. Let $p_1 = p \wedge \hat{x}_0$ be the 'infimum' of the sublinear functionals p and $\hat{x}_0$ on C' i.e.,

$$p_1(f) = \inf\{p(f_1) + \hat{x}_0(f_2) : f = f_1 + f_2, f_1, f_2 \in C'\} \qquad (f \in C').$$

Similarly let $q_1 = q \vee \left(\hat{x}_0 - \frac{\|\cdot\|}{2}\right)$ be the 'supremum' of the superlinear

functionals q and $\left(\hat{x}_0 - \frac{\|\cdot\|}{2}\right)$ i.e.,

$$q_1(f) = \sup\left\{q(f_1) + \hat{x}_0(f_2) - \frac{\|f_2\|}{2} : f = f_1 + f_2, f_1, f_2 \in C'\right\} \qquad (f \in C').$$

Then p_1 is lower semi-continuous and q_1 is upper semi-continuous on $C' \cap \Sigma'$ (and hence on C' by the Krein–Smulian theorem) with respect to the $\sigma(E', E)$-topology. Furthermore, $q_1 \leqslant p_1$ on C'. In fact, let $f = f_1 + f_2 = g_1 + g_2 \in C'$, where f_1, f_2, g_1, g_2 are in C'. Since (E', C') has the Riesz decomposition property, there exist $h_{11}, h_{12}, h_{21}, h_{22}$ in C' such that

$$f_1 = h_{11} + h_{21}, \qquad f_2 = h_{12} + h_{22}$$
$$g_1 = h_{11} + h_{12}, \qquad g_2 = h_{21} + h_{22}.$$

Since $\|\cdot\|$ is a Riesz norm, $\|h_{12}\| \leqslant \|f_2\|$. By the linearity of p, q, it follows from inequality (9.4) that

$$q(f_1) + \hat{x}_0(f_2) - \frac{\|f_2\|}{2} \leqslant q(h_{11}) + q(h_{21}) + \hat{x}_0(h_{12}) + \hat{x}_0(h_{22}) - \frac{\|h_{12}\|}{2}$$
$$\leqslant p(h_{11}) + \hat{x}_0(h_{21}) + p(h_{12}) + \hat{x}_0(h_{22})$$
$$= p(h_{11} + h_{12}) + \hat{x}_0(h_{21} + h_{22})$$
$$= p(g_1) + \hat{x}_0(g_2).$$

By the definition of q_1 and p_1, it follows that $q_1(f) \leqslant p_1(f)$ for all $f \in C'$. Applying corollary (2.19) again, there is x_1 in E such that

$$q_1(f) \leqslant f(x_1) \leqslant p_1(f) + \frac{\|f\|}{2^2} \qquad (f \in C').$$

Then

$$q(f) \leqslant f(x_1) \leqslant p(f) + \frac{\|f\|}{2^2} \qquad (f \in C')$$

and

$$-\frac{\|f\|}{2} \leqslant f(x_1) - f(x_0) \leqslant \frac{\|f\|}{2^2} \qquad (f \in C').$$

Inductively, we can construct a sequence $\{x_n\}$ in E such that

$$q(f) \leqslant f(x_n) \leqslant p(f) + \frac{\|f\|}{2^{n+1}} \qquad (f \in C')$$

and

$$-\frac{\|f\|}{2^{n+1}} \leqslant f(x_{n+1} - x_n) \leqslant \frac{\|f\|}{2^{n+2}} \qquad (f \in C')$$

for each positive integer n. Since the norm $\|\,.\,\|$ on E' is a Riesz norm, it follows easily from the Hahn–Banach theorem and the last displayed inequalities that $\{x_n\}$ is a Cauchy sequence in the Banach space $(E, \|\,.\,\|)$, and hence converges, say, to x in E. For this x, we clearly have

$$q(f) \leqslant f(x) \leqslant p(f) \qquad (f \in C'),$$

as required.

The preceding separation theorem enables us to establish the following important duality theorem.

(9.20) THEOREM (Riesz–Andô). *Let $(E, C, \|\,.\,\|)$ be an ordered Banach space with closed cone C, and suppose that C is normal and generating in $(E, \|\,.\,\|)$. Let $(E', C', \|\,.\,\|)$ be the Banach dual space with the dual cone C'. Then the following statements are equivalent:*

(a) *(E, C) has the Riesz decomposition property;*

(b) *(E', C') is a vector lattice;*

(c) *(E', C') has the Riesz decomposition property.*

Proof. By corollary (8.3), (a) $\Rightarrow$ (b), and the implication (b) $\Rightarrow$ (c) is elementary and well known. Thus it remains to show (c) $\Rightarrow$ (a). Suppose (c) holds and that $x_1, x_2 \leqslant y_1, y_2$ in E. Let

$$q = \hat{x}_1 \vee \hat{x}_2 \quad \text{and} \quad p = \hat{y}_1 \wedge \hat{y}_2,$$

where $\hat{x}_i$ and $\hat{y}_i$ $(i = 1, 2)$ are the canonical images in E'' of x_i and y_i respectively. Then p, q satisfy the conditions of the preceding theorem, so there exists x in X such that

$$q(f) \leqslant f(x) \leqslant p(f) \qquad (f \in C').$$

Then $f(x_1) \leqslant q(f) \leqslant f(x)$ for all $f \in C'$. Since C is closed, it follows that $x_1 \leqslant x$. Similarly we can show $x_2 \leqslant x$ and $x \leqslant y_1, y_2$. This shows that (a) holds.

(9.21) COROLLARY. *Let $(E, C, \|\,.\,\|)$ and $(E', C', \|\,.\,\|)$ be as in the preceding theorem. Then the following statements are equivalent:*

(a) *$(E', C', \|\,.\,\|)$ is a Banach lattice;*

(b) (E, C) has the Riesz decomposition property and the norm $\|\,.\,\|$ on E is a Riesz norm.

In particular, the Banach dual of a Banach lattice is a Banach lattice.

Proof. Follows from theorems (6.12) and (9.20).

The following result was proved independently by Davies (1967) and Ng (unpublished) at about the same time.

(9.22) COROLLARY. Let $(E, C, \|\,.\,\|)$ and $(E', C', \|\,.\,\|)$ be as in the preceding theorem. Then the following statements are equivalent:

(a) $(E', C', \|\,.\,\|)$ is an AL-space;

(b) $(E, C, \|\,.\,\|)$ is an approximate order-unit normed space and has the Riesz decomposition property.

In particular, if E is an AM-space then E' must be an AL-space.

Proof. By proposition (9.18), (a) holds if and only if (E', C') is a vector lattice and the norm on E' is a base norm. Thus the equivalence of (a) and (b) follows immediately from theorems (9.20) and (9.9).

Remark. It may happen that (E, C) is not a vector lattice even though $(E', C', \|\,.\,\|)$ is an AL-space. For a counterexample, see Lindenstrass (1964). The following result of Ellis (1964) is dual to corollary (9.22).

(9.23) COROLLARY. Let $(E, C, \|\,.\,\|)$ be an ordered Banach space with closed cone C, and suppose that C is normal and generating in $(E, \|\,.\,\|)$. Let $(E', C', \|\,.\,\|)$ be the Banach dual space with the dual cone C'. Then the following statements are equivalent:

(a) $(E, C, \|\,.\,\|)$ is an AL-space;

(b) $(E', C', \|\,.\,\|)$ is an AM-space;

(c) (E, C) has the Riesz decomposition property, and the norm $\|\,.\,\|$ on E is a Riesz norm and additive on C;

(d) (E, C) has the Riesz decomposition property, and $(E, C, \|\,.\,\|)$ is a base normed space.

Proof. By proposition (9.5), (c) $\Leftrightarrow$ (d). By proposition (9.17), (b) holds if and only if E' is a vector lattice and the norm on E' is an approximate order-unit norm. Thus the equivalence (b) and (d) follows from theorems (9.20) and (9.8). Therefore statements (b), (c), and (d) are mutually equivalent. Further, it is trivial that (a) $\Rightarrow$ (c). Thus to complete the proof, we have only to show (b) $\Rightarrow$ (a). Accordingly, suppose (b) holds. Then (c) holds and the second dual space $(E'', C'', \|\cdot\|)$ is an AL-space by corollary (9.22). By virtue of (c) and proposition (9.18), to show that E is an AL-space it is sufficient to show that it is a vector lattice. Now, let $x \in E$. By (c), for each positive integer n, there exists x_n in E with $\|x_n\| < \|x\| + 1/n$ such that $\pm x \leqslant x_n$. Let

$$y_n = \tfrac{1}{2}(x_n + x) \quad \text{and} \quad z_n = \tfrac{1}{2}(x_n - x).$$

Then $y_n, z_n \geqslant 0$, $x_n = y_n + z_n$ and $x = y_n - z_n$. Let $\hat{y}_n, \hat{z}_n$, and $\hat{x}$ respectively denote the canonical images of y_n, z_n, and x in the AL-space E''. Let $\hat{x}^+ = \hat{x} \vee 0$ and $\hat{x}^- = -\hat{x} \vee 0$ in the vector lattice E''. Then $\hat{y}_n \geqslant \hat{x}^+$ and $\hat{z}_n \geqslant \hat{x}^-$. Hence, since E'' is an AL-space, we have

$$\|y_n\| = \|\hat{y}_n\| = \|\hat{y}_n - \hat{x}^+\| + \|\hat{x}^+\|$$

and

$$\|z_n\| = \|\hat{z}_n\| = \|\hat{z}_n - \hat{x}^-\| + \|\hat{x}^-\|.$$

Consequently

$$\|x\| + \frac{1}{n} > \|x_n\| = \|y_n + z_n\| = \|y_n\| + \|z_n\| = \|\hat{y}_n - \hat{x}^+\|$$
$$+ \|\hat{z}_n - \hat{x}^-\| + \|\hat{x}^+\| + \|\hat{x}^-\| = \|\hat{y}_n - \hat{x}^+\| + \|\hat{z}_n - \hat{x}^-\| + \|\hat{x}^+ + \hat{x}^-\|$$
$$= \|\hat{y}_n - \hat{x}^+\| + \|\hat{z}_n - \hat{x}^-\| + \|\hat{x}\| = \|\hat{y}_n - \hat{x}^+\| + \|\hat{z}_n - \hat{x}^-\| + \|x\|.$$

Passing to the limit, as $n \to \infty$ we see that $\hat{y}_n \to \hat{x}^+$ and $\hat{z}_n \to \hat{x}^-$. Since $(E, \|\cdot\|)$ is complete, its image $\hat{E}$ in E'' must be closed; hence $\hat{x}^+, \hat{x}^- \in \hat{E}$. This shows that $\hat{E}$ must be a vector sublattice of E''; hence E is a vector lattice in its own right. This completes the proof of corollary (9.23).

Recall that a Banach lattice $(E, C, \|\cdot\|)$ is an AL- or (AL^1-) space if the norm is additive on E, and that it is an AM- (or AL^∞-) space if the open unit ball is directed upwards. Thus, in a general ordered normed space, the additivity of the norm on the cone may be called an AL^1-condition and the directedness of the open unit ball may be called an AL^∞-condition. In view of theorems (9.6) and (9.7), these two conditions are dual. In the following we introduce what may be called L^p-($1 < p < \infty$) conditions and study some duality problems involving such conditions. For simplicity we shall discuss the case when $(E, \|\cdot\|)$ *is a Banach space and C is a* $\|\cdot\|$*-closed cone in* $(E, \|\cdot\|)$. Except where we state to the contrary, p and q will denote positive real numbers such that $1/p + 1/q = 1$. In $(E, C, \|\cdot\|)$, we consider the following conditions linking the norm and the ordering.

L^p-*condition* (*i*): if $x, y \in C$ then $\|x + y\|^p \geqslant \|x\|^p + \|y\|^p$.

L^p-*condition* (*ii*): if $x, y \in E$ and $\varepsilon > 0$ then there exists $z \in E$ with $\|z\|^p \leqslant \|x\|^p + \|y\|^p + \varepsilon$ such that $z \geqslant x, y$.

Let $(E', C', \|.\|)$ denote the ordered Banach dual space with the dual cone C'. We shall show that the L^p-condition (i) and the L^p-condition (ii) are dual conditions.

(9.24) LEMMA. *For any ordered Banach space $(E, C, \|.\|)$ with closed cone C, E satisfies L^p-condition (ii) if and only if E' satisfies L^q-condition (i).*

Proof. (a) *Necessity.* Let $f, g \in E'$ be such that $f, g \geqslant 0$. We have to show that

$$\|f+g\|^q \geqslant \|f\|^q + \|g\|^q. \tag{9.5}$$

If $f = 0$ or $g = 0$, then inequality (9.5) is trivial. We may therefore suppose that $f \neq 0$, $g \neq 0$. Take a real number such that

$$0 < \varepsilon < \|f\|, \|g\|.$$

Then there exist u, v in E with $\|u\| < 1$ and $\|v\| < 1$ such that

$$\|f\| - \varepsilon < f(u) \quad \text{and} \quad \|g\| - \varepsilon < g(v).$$

Let
$$x = (\|f\| - \varepsilon)^{q/p} u \quad \text{and} \quad y = (\|g\| - \varepsilon)^{p/q} v.$$

Since E satisfies L^p-condition (ii), there exists $z \in E$ such that $z \geqslant x, y$ and

$$\|z\|^p \leqslant \|x\|^p + \|y\|^p + \varepsilon < (\|f\| - \varepsilon)^q + (\|g\| - \varepsilon)^q + \varepsilon.$$

Since f, g are positive and $q = 1 + q/p$, we have

$$(\|f\| - \varepsilon)^q + (\|g\| - \varepsilon)^q \leqslant f(x) + g(y) \leqslant f(z) + g(z) \leqslant \|f+g\| \|z\|,$$

and it follows that

$$\frac{(\|f\| - \varepsilon)^q + (\|g\| - \varepsilon)^q}{\{(\|f\| - \varepsilon)^q + (\|g\| - \varepsilon)^q + \varepsilon\}^{1/p}} \leqslant \|f+g\|.$$

Passing to the limit as $\varepsilon \to 0$, we have

$$(\|f\|^q + \|g\|^q)^{1/q} = (\|f\|^q + \|g\|^q)^{1-1/p} \leqslant \|f+g\|,$$

proving inequality (9.5).

(b) *Sufficiency.* Let x, y be in E and $\varepsilon > 0$. Define

$$Q(h) = \sup\{f(x) + g(y) : f, g \in C', h = f+g\} \quad (h \in C')$$

and
$$P(h) = \|h\| (\|x\|^p + \|y\|^p)^{1/p} \quad (h \in E').$$

Then Q is an upper $\sigma(E', E)$-semi-continuous real-valued superlinear functional such that $Q(h) \geqslant h(x), h(y)$ for all $h \in C'$; and P is a lower $\sigma(E', E)$-semi-continuous real-valued sublinear functional. Also, since

E' satisfies L^q-condition (i), by Hölder's inequality we have

$$f(x)+g(y) \leqslant \|f\| \|x\| + \|g\| \|y\| \leqslant (\|f\|^q + \|g\|^q)^{1/q}(\|x\|^p + \|y\|^p)^{1/p}$$
$$\leqslant \|f+g\| (\|x\|^p + \|y\|^p)^{1/p} = P(h),$$

whenever $f+g = h$ and f, g, h are in C'. This shows that $Q(h) \leqslant P(h)$ for all $h \in C'$. Since E is complete, by a standard Hahn–Banach separation argument (cf. corollary (2.19)), we can find $z \in E$ such that

$$Q(h) \leqslant h(z) \qquad (h \in C'),$$
$$h(z) \leqslant P(h)+\delta \|h\| \qquad (h \in E'),$$

where δ is a positive real number such that

$$\{(\|x\|^p + \|y\|^p)^{1/p} + \delta\}^p < \|x\|^p + \|y\|^p + \varepsilon.$$

Then $\|z\|^p \leqslant \|x\|^p + \|y\|^p + \varepsilon$. Also, since C is closed, it follows from

$$h(x) \leqslant Q(h) \leqslant h(z) \qquad (h \in C')$$

that $x \leqslant z$. Similarly $y \leqslant z$.

(9.25) LEMMA. *Let* $(E, C, \|.\|)$ *be as in the preceding lemma. Then E satisfies L^p-condition (i) if and only if E' satisfies L^q-condition (ii), i.e. if and only if E' satisfies the following condition:*

(iii) *if f, g are in E', then there exists $h \in E'$ with $\|h\|^q \leqslant \|f\|^q + \|g\|^q$ such that $h \geqslant f, g$.*

Proof. We first remark that in the situation of Banach dual spaces, the ε that appeared in L^p-condition (ii) can be dropped (by a compactness argument).

If E' satisfies L^q-condition (ii), then it follows from the necessity part of lemma (9.24) that E'' satisfies L^p-condition (i), so does E, since E is isometrically isomorphic to a subspace of E'' (under the canonical embedding). Conversely, suppose that E satisfies L^p-condition (i), and let $f, g \in E'$. Define

$$Q(z) = \sup\{f(x)+g(y) : x, y \in C, z = x+y\} \qquad (z \in C)$$

and

$$P(z) = \|z\| (\|f\|^q + \|g\|^q)^{1/q} \qquad (z \in E).$$

Then P and $-Q$ are sublinear functionals on where they are defined. Also, since E satisfies L^p-condition (i), by Hölder's inequality, we can show that $Q(z) \leqslant P(z)$ for all $z \in C$. By Bonsall's generalization of the Hahn–Banach theorem (1.15), there exists a linear functional h on E

such that
$$Q(w) \leqslant h(w) \quad \text{and} \quad h(z) \leqslant P(z)$$

for all $w \in C$ and $z \in E$. Then $h \in E'$ and satisfies the required property in (iii), and *a fortiori* E' satisfies L^q-condition (ii).

Remark. From our proofs it is clear that the preceding lemma and the necessity part of lemma (9.24) are still valid even if E and C are not complete.

An ordered Banach space (not necessarily a vector lattice) is said to satisfy the L^p-*conditions* if it satisfies both L^p-conditions (i) and (ii). Combining lemmas (9.24) and (9.25) we arrive at the following theorem.

(9.26) THEOREM. *Let p, q be real numbers such that $1/p + 1/q = 1$, and $(E, C, \|.\|)$ an ordered Banach space with a closed cone C. Then E satisfies L^p-conditions if and only if the ordered Banach dual space E' satisfies L^q-conditions.*

It should be noted that an ordered Banach $(E, C, \|.\|)$ is an AL^p-*space* if it is a Banach lattice and satisfies the L^p-conditions. Notice that in an AL^p-space E, the following equalities hold:

$$\|x\|^p = \| \, |x| \, \|^p = \|x^+\|^p + \|x^-\|^p \qquad (x \in E),$$

where $|x| = x \vee -x$, $x^+ = x \vee 0$ and $x^- = -x \vee 0$.

(9.27) THEOREM. *Let p, q be real numbers such that $1/p + 1/q = 1$, and $(E, C, \|.\|)$ an ordered Banach space with closed cone C. Then E is an AL^p-space if and only if E' is an AL^q-space.*

Proof. If E is an AL^p-space, then E' is a Banach lattice (by theorem (6.12)), and satisfies AL^q-conditions (by the preceding theorem); hence E' is an AL^q-space. Conversely, suppose that E' is an AL^q-space. Then E'' is an AL^p-space and E satisfies AL^p-conditions by theorem (9.26). Further, by corollary (9.21), (E, C) has the Riesz decomposition property, and the norm $\|.\|$ on E is a Riesz norm. Thus, to show that E is an AL^p-space, it remains to show that E is a vector lattice. We prove this by a similar argument as that given in corollary (9.23). Let x be in E. For each positive integer n, there exists $x_n \in E$ with $\|x_n\| < \|x\| + \dfrac{1}{n}$ such that $\pm x \leqslant x_n$. Suppose $y_n = \frac{1}{2}(x_n + x)$ and $z_n = \frac{1}{2}(x_n - x)$. Then

$y_n, z_n \geqslant 0$ and $x = y_n - z_n$. Since E satisfies L^p-conditions, it follows that

$$\|y_n\|^p + \|z_n\|^p \leqslant \|y_n + z_n\|^p = \|x_n\|^p < \left(\|x\| + \frac{1}{n}\right)^p. \tag{9.6}$$

On the other hand, let ϕ denote the canonical embedding of E into E'' and let $\phi(x)^+ = \phi(x) \vee 0$ and $\phi(x)^- = -\phi(x) \vee 0$ in the vector lattice E''. Since E'' is an AL^p-space, we have

$$\|\phi(x)^+\|^p + \|\phi(x)^-\|^p = \|\phi(x)\|^p = \|x\|^p.$$

Notice also that $\phi(y_n) \geqslant \phi(x)^+$ and $\phi(z_n) \geqslant \phi(x)^-$. Hence

$$\|y_n\|^p = \|\phi(y_n)\|^p = \|\phi(y_n) - \phi(x)^+ + \phi(x)^+\|^p$$
$$\geqslant \|\phi(y_n) - \phi(x)^+\|^p + \|\phi(x)^+\|^p$$

and

$$\|z_n\|^p = \|\phi(z_n)\|^p = \|\phi(z_n) - \phi(x)^- + \phi(x)^-\|^p$$
$$\geqslant \|\phi(z_n) - \phi(x)^-\|^p + \|\phi(x)^-\|^p,$$

so it follows from formula (9.6) that

$$\left(\|x\| + \frac{1}{n}\right)^p \geqslant \|y_n\|^p + \|z_n\|^p \geqslant \|\phi(y_n) - \phi(x)^+\|^p + \|\phi(z_n) - \phi(x)^-\|^p + \|x\|^p.$$

Passing to the limit as $n \to \infty$, we see that $\phi(y_n) \to \phi(x)^+$ and $\phi(z_n) \to \phi(x)^-$. Since E is complete, it follows that $\phi(x)^+ \in \phi(E)$ and $\phi(x)^- \in \phi(E)$. Therefore $\phi(E)$ is a vector sublattice of E'', and so E is a vector lattice. Consequently E is an AL^p-space.

Remark. As in theorem (9.26), the proof of the necessity part does not require the assumption of the completeness of E and C.

(9.28) THEOREM. *Let $(E, C, \|.\|)$ be an ordered Banach space with closed cone C. Suppose that the norm $\|.\|$ is a Riesz norm on (E, C) and that $(E, C, \|.\|)$ satisfies AL^p-conditions $(1 \leqslant p < \infty)$. Then E has the Riesz decomposition property if and only if it is an AL^p-space.*

Proof. Let q be the (extended) real number such that $1/p + 1/q = 1$ (if $p = 1$ then $q = \infty$). We only have to show the necessity part. Accordingly, suppose that E has the Riesz decomposition property and satisfies the assumptions of the theorem. Then the ordered Banach dual space E' is a Banach lattice (by corollary (9.21)) and satisfies the AL^q-conditions (by theorems (9.26) and (9.6)); hence E' is an AL^q-space. From the preceding theorem (and corollary (9.23)), we conclude that E is an AL^p-space.

10

ELEMENTARY THEORY OF RIESZ SPACES

WE recall that an ordered vector space (X, C) with a proper cone C is called a *Riesz space* (or *vector lattice*) if each pair of elements x, y of X has a *least upper bound*, written $x \vee y$ (or $\sup(x, y)$), in X. Equivalently, (X, C) is a Riesz space if and only if each pair of elements x, y of X has a *greatest lower bound*, written $x \wedge y$ (or $\inf(x, y)$), in X. Indeed, $x \vee y = -(-x \wedge -y)$ if one of them exists.

From now on (X, C) (or simply X) will denote a Riesz space with the positive cone C. It is clear that each Riesz space must be a weakly Riesz space. The following example shows that there are weakly Riesz spaces which are not Riesz spaces. Consider R^2 with the cone defined by

$$C_{\mathrm{w}} = \{(x_1, x_2) \in R^2 : x_1 > 0, x_2 > 0\} \cup \{(0, 0)\}.$$

Then (R^2, C_w) is a weakly Riesz space but it is not a Riesz space. For further examples, see Fuchs (1966).

(10.1) PROPOSITION. *Let (X, C) be a Riesz space. Then the following statements hold:*

(a) $(x \vee y) + z = (x+z) \vee (y+z)$ *and* $(x \wedge y) + z = (x+z) \wedge (y+z)$;

(b) *if u, v, w are in C then* $(u+v) \wedge w \leqslant (u \wedge w) + (v \wedge w)$;

(c) *if $x \wedge z = y \wedge z = 0$ then* $(x+y) \wedge z = 0$;

(d) *if $u = x+y = z+w$ then* $u = (x \vee z) + (y \wedge w)$;

(e) $x+y = x \vee y + x \wedge y$.

Proof. It is straightforward to verify (a); (c) follows easily from (b) since x, y, z are certainly in C; (e) follows from (d) (by putting $z = y$ and $w = x$). To prove (b), let $z = (u+v) \wedge w$. Then $0 \leqslant z \leqslant u+v$, we have by the Riesz decomposition property that $z = x+y$, where $0 \leqslant x \leqslant u$, $0 \leqslant y \leqslant u$. On the other hand, since $x \leqslant z \leqslant w, y \leqslant z \leqslant w$, it follows that $x \leqslant u \wedge w$ and $y \leqslant u \wedge w$; these imply that

$$(u+v) \wedge w = z \leqslant u \wedge w + v \wedge w.$$

Finally, note that the statement (d) is equivalent to

$$y - (y \wedge w) = (x \vee z) - x. \tag{10.1}$$

Notice also that

$$y-(y \wedge w) = y+(-y \vee -w) = 0 \vee (y-w),$$
$$x \vee z-x = (x-x) \vee (z-x) = 0 \vee (z-x).$$

By hypothesis $y-w = z-x$, thus equality (10.1) is clear.

(10.2) PROPOSITION. *A Riesz space (X, C) is distributive, that is:*
  (a) *if $\sup\{x_\alpha : \alpha \in I\}$ exists in X then*

$$y \wedge \sup\{x_\alpha : \alpha \in I\} = \sup\{y \wedge x_\alpha : \alpha \in I\},$$

  (b) *if $\inf\{x_\alpha : \alpha \in I\}$ exists in X then*

$$y \vee \inf\{x_\alpha : \alpha \in I\} = \inf\{y \vee x_\alpha : \alpha \in I\}.$$

Proof. Let $x = \sup\{x_\alpha : \alpha \in I\}$. Then it is obvious that $y \wedge x \geqslant y \wedge x_\alpha$ for all $\alpha \in I$. Suppose that $z \geqslant y \wedge x_\alpha$ for all $\alpha \in I$. To prove (a), we have to show that $z \geqslant y \wedge x$. In view of proposition (10.1)(e), it is equivalent to verify that $z \geqslant x+y-(y \vee x)$. Now, for each $\alpha \in I$, we have

$$z+(y \vee x) \geqslant z+(y \vee x_\alpha) \geqslant y \wedge x_\alpha+y \vee x_\alpha = y+x_\alpha.$$

Consequently, $$z+(y \vee x) \geqslant y+x,$$

as required. The proof of (b) is similar and will be omitted.

If (X, C) is a Riesz space and if x is in X, we define

$$x^+ = x \vee 0 \qquad x^- = (-x) \vee 0 \quad \text{and} \quad |x| = x \vee (-x).$$

x^+ and x^- are called the *positive part* and the *negative part*, respectively, of the element x, while $|x|$ is referred to as the *absolute value* of x. It is easily seen from proposition (10.1)(e) that $x = x^+-x^-$. Two elements x and y of X are said to be *disjoint*, written $x \perp y$, if $|x| \wedge |y| = 0$. For any subset B of X, we write

$$B^d = \{x \in X : x \perp b \quad \text{for all } b \text{ in } B\}.$$

(10.3) PROPOSITION. *Elements of a Riesz space X satisfy the following properties:*
  (a) $x^+ \perp x^-$;
  (b) $|x| = x^++x^- = x^+ \vee x^-$;

(c) $x = x^+ - x^-$ is the unique representation of x as a difference of two disjoint positive elements;

(d) $(x+y)^+ \leqslant x^+ + y^+$ and $(x+y)^- \leqslant x^- + y^-$;

(e) $|x+y| \leqslant |x| + |y|$;

(f) $|x-y| = x \vee y - x \wedge y = |x \vee z - y \vee z| + |x \wedge z - y \wedge z|$;

(g) $|x^+ - y^+| \leqslant |x-y|$ and $|x^- - y^-| \leqslant |x-y|$;

(h) $x \perp y$ if and only if $|x| + |y| = |\,|x| - |y|\,|$;

(i) $|x| \leqslant u$ if and only if $-u \leqslant x \leqslant u$.

Proof. (a) Observe that

$$x^+ \wedge x^- - x^- = (x^+ - x^-) \wedge 0 = x \wedge 0 = -x^-,$$

it then follows that $x^+ \perp x^-$.

(b) In view of proposition (10.1)(a) and what we have just proved, we have

$$x^+ + x^- = x^+ \vee x^- = x \vee 0 \vee ((-x) \vee 0) = x \vee (-x) = |x|.$$

(c) Suppose that $x = u - w$, where $u, w \geqslant 0$ and $u \perp w$. It is required to show that $u = x^+$ and $w = x^-$. Notice that

$$0 \leqslant u - x^+ \leqslant u, \qquad 0 \leqslant w - x^- \leqslant w, \qquad u - x^+ = w - x^-,$$

it follows from $u \perp w$ that $w - x^- = u - x^+ = (u - x^+) \wedge (w - x^-) = 0$, and hence that $u = x^+$, $w = x^-$.

(d) Clearly $0, x+y \leqslant x^+ + y^+$; hence $(x+y)^+ \leqslant x^+ + y^+$. Similarly we can show that $(x+y)^- \leqslant x^- + y^-$.

(e) In view of (b) and (d) of this proposition, we have

$$|x+y| = (x+y)^+ + (x+y)^- \leqslant x^+ + y^+ + x^- + y^- = |x| + |y|.$$

(f) Observe, for any x, y in X, that

$$|x-y| = (x-y)^+ + (x-y)^- = (x-y) \vee 0 - (x-y) \wedge 0$$
$$= x \vee y - y - (x \wedge y - y) = x \vee y - x \wedge y,$$

we then have, by proposition (10.1)(e), that

$$|x \vee z - y \vee z| + |x \wedge z - y \wedge z| = (x \vee z) \vee (y \vee z) - (x \vee z) \wedge (y \vee z)$$
$$+ (x \wedge z) \vee (y \wedge z) - (x \wedge z) \wedge (y \wedge z)$$
$$= x \vee y \vee z - (x \wedge y) \vee z + (x \vee y) \wedge z - x \wedge y \wedge z$$
$$= (x \vee y + z) - (x \wedge y + z) = x \vee y - x \wedge y = |x-y|.$$

(g) In view of (f) of this proposition, we have

$$|x^+-y^+| \leqslant |x^+-y^+|+|x \wedge 0-y \wedge 0| = |x-y|$$
$$|x^--y^-| \leqslant |x^+-y^+|+|-x^-+y^-| = |x-y|.$$

(h) By proposition (10.1)(c) and (f) of this proposition, we see that

$$|x|+|y| = |x| \vee |y|+|x| \wedge |y| = \big||x|-|y|\big|+2\,|x| \wedge |y|.$$

Recall that $x \perp y$ if and only if $|x| \wedge |y| = 0$, thus this is the case if and only if

$$|x|+|y| = \big||x|-|y|\big|.$$

Finally, it is trivial to verify (i).

We recall that a set S in an ordered vector space E is *solid* if and only if $S = \cup\{[-u, u]:0 < u \in S\}$. In terms of lattice structure, we are able to give some characterization of solid sets as follows.

(10.4) PROPOSITION. *A set S in a Riesz space X is solid if and only if it satisfies the following property:*

$$|x| < |y| \quad with \quad y \in S \Rightarrow x \in S. \tag{10.2}$$

Proof. If S is solid and if $|x| < |y|$ with $y \in S$, then there exists $0 < u \in S$ such that $-u \leqslant y \leqslant u$ and so, by proposition (10.3)(i), $|y| \leqslant u$. Observe that $|x| \leqslant u$; apply proposition (10.3)(i) again, $-u \leqslant x \leqslant u$, consequently $x \in S$. Conversely suppose statement (10.2) holds. Then $|x| \in S$ whenever $x \in S$; it follows from $-|x| \leqslant x \leqslant |x|$ that $S \subset \cup\{[-u, u]:0 < u \in S\}$. On the other hand, if $-u \leqslant x \leqslant u$ for some $0 < u \in S$ then, by proposition (10.3)(i), $|x| \leqslant u$, and hence $x \in S$. Therefore $S = \cup\{[-u, u]:0 < u \in S\}$, and the proof is complete.

It is easily seen that the intersection of a family of solid sets in a Riesz space X is either empty or solid, and that the union of a family of solid sets in X is solid. If B is a subset of X, the smallest solid set containing B, written S_B, is called the *solid hull* of B. It is clear that

$$S_B = \cup\{[-|b|, |b|]:b \in B\}.$$

The solid hull of an element x in X will be denoted by S_x; therefore $S_x = [-|x|, |x|]$. If B is a subset of X, the set defined by

$$\text{sk}(B) = \{x \in X:[-|x|, |x|] \subset B\}$$

is called the *solid kernel* of B. It is clear that sk(B) is either empty or the largest solid subset of X contained in B, and that

$$\text{sk}(B) = \cup\{[-u, u]:[-u, u] \subseteq B\}.$$

We recall that a set B in X is absolute-order-convex if and only if $S(B) \subset B$ and absolutely dominated if $B \subset S(B)$, where

$$S(B) = \cup\{[-u, u]:0 \leqslant u \in B\}.$$

Therefore, if B is absolute-order-convex then $S(B)$ is the solid kernel of B, i.e. $S(B) = \mathrm{sk}(B)$. If B is absolutely dominated then $S(B)$ is the solid hull of B, i.e. $S(B) = S_B$.

Some elementary, but useful, properties of solid sets are summarized in the following proposition.

(10.5) PROPOSITION. *Let (X, C) be a Riesz space, and let V be a subset of X. Then the following statements hold:*

(a) *the convex hull of each solid set in X is solid;*

(b) *if V is convex then so is $\mathrm{sk}(V)$;*

(c) *if $\mathrm{sk}(V)$ is non-empty then $\mathrm{sk}(V)$ is absorbing if and only if $\mathrm{sk}(V)$ absorbs every order-bounded subset of X;*

(d) *V absorbs every order-bounded set in X if and only if $\mathrm{sk}(V)$ absorbs all order-bounded subsets of X;*

(e) *if A, B are solid subsets of X then $A+B$ and λB are solid for any real number λ.*

Proof. (a) It is clear that $S_{\lambda x} = \lambda S_x$ for any real number λ and $x \in X$. We now claim that $S_{x+y} \subseteq S_x + S_y$. If $z \in S_{x+y}$, then

$$|z| \leqslant |x+y| \leqslant |x|+|y|$$

and so $0 \leqslant (z+|x|+|y|)/2 \leqslant |x|+|y|$. By the Riesz decomposition property, there exist w_1 and w_2 with $0 \leqslant w_1 \leqslant |x|$, $0 \leqslant w_2 \leqslant |y|$ such that

$$z+|x|+|y| = 2w_1+2w_2.$$

Take $u = 2w_1-|x|$, $v = 2w_2-|y|$; then $u \in S_x$, $v \in S_y$ and $z = u+v$; this implies that $S_{x+y} \subset S_x + S_y$.

If A is a solid subset of X and if $x \in \mathrm{co}\, A$, there exist $a_i \in A$ and $\lambda_i \in [0, 1]$ with $\sum_{i=1}^{n} \lambda_i = 1$ such that $x = \sum_{i=1}^{n} \lambda_i a_i$. We now conclude, from

$$S_x \subseteq \sum_{i=1}^{n} \lambda_i S_{a_i} \subseteq \sum_{i=1}^{n} \lambda_i A \subseteq \mathrm{co}\, A,$$

that $\mathrm{co}\, A$ is solid.

(b) Notice that $\mathrm{sk}(V)$ is the largest solid subset of X contained in V, and that $\mathrm{co}(\mathrm{sk}(V))$ is solid. We conclude from $\mathrm{co}(\mathrm{sk}(V)) \subseteq V$ that $\mathrm{sk}(V) = \mathrm{co}(\mathrm{sk}(V))$, and hence that $\mathrm{sk}(V)$ is convex.

9

(c) Since sk(V) is solid, sk(V) absorbs $u \in C$ if and only if it absorbs $[-u, u]$.

(d) The condition is clearly sufficient. To prove its necessity, it is sufficient to show that sk(V) absorbs each order-interval of the form $[-u, u]$ where $u \in C$. Since V absorbs $[-u, u]$, there exists $\lambda > 0$ such that $[-u, u] \subset \lambda V$, and so $[-u, u] \subset$ sk(λV) $= \lambda$ sk(V).

(e) It is clear that λB is solid. It remains to verify that $A + B$ is solid. Let $x \in A$, $y \in B$, and let z, in X, be such that $|z| \leqslant |x + y|$. It is known from the proof of (a) that $S_{x+y} \subseteq S_x + S_y$; hence there exist $a \in S_x$ and $b \in S_y$ such that $z = a + b$. Since A and B are solid, it follows that $S_x \subset A$, $S_y \subset B$, and hence that $z \in A + B$. Therefore $A + B$ is solid, and the proof of this proposition is complete.

It should be noted that the solid hull of a convex set in a Riesz space is, in general, not convex. By way of example, consider $\mathbf{R}^2$ with a positive cone C defined by

$$C = \{(x, y) : x \geqslant 0, y \geqslant 0\},$$

then $(\mathbf{R}^2, C)$ is a Riesz space. Suppose that

$$B = \{\lambda(-2, 0) + (1 - \lambda)(1, 3) : \lambda \in [0, 1]\}.$$

Then B is convex, but the solid hull S_B of B is not convex because $\{\lambda(-2, 0) + (1 - \lambda)(-1, 3) : \lambda \in [0, 1]\}$ is not contained in S_B.

A vector subspace S of a Riesz space (X, C) is called a *Riesz subspace* if $x^+ \in S$ whenever $x \in S$. Solid subspaces of X are referred to as *lattice-ideals* (or simply *ℓ-ideals*). An *ℓ*-ideal B of X is called a *normal subspace* of X if it follows from $x_\tau \uparrow x$ in X with x_τ in B for all τ that x belongs to B. It is clear that the intersection of a family of normal subspaces of X is a normal subspace. An *ℓ*-ideal B of X is called a *σ-normal subspace* of X if it follows from $x_n \uparrow x$ in X with x_n in B for all n that x belongs to B. The intersection of a family of *σ*-normal subspaces of X is a *σ*-normal subspace, of *ℓ*-ideals in X is an *ℓ*-ideal and of Riesz subspaces of X is a Riesz subspace. If B is a subset of X, the smallest *ℓ*-ideal containing B is called the *ℓ-ideal generated by B*. If B is a subset of X, the smallest Riesz subspace containing B is called the *Riesz subspace generated by B*. It is clear that if B is a Riesz subspace of X, then the order-convex hull $[B]$ of B is the *ℓ*-ideal in X generated by B. If S is an *ℓ*-ideal in X, the smallest normal subspace containing S, written $\{S\}$, is called the *normal subspace generated by S*. If S is an *ℓ*-ideal in X, the smallest *σ*-normal subspace containing S, written

$\{S\}_\sigma$, is called the *σ-normal subspace generated by* S. For instance, if A is a subset of (X, C), then A^{d} is a normal subspace of X. If S and B are ℓ-ideals in X, S is said to be *order-dense* in B if $B \subset \{S\}$; S is *σ-order-dense* if $B \subset \{S\}_\sigma$. In particular, S is said to be *order-dense* if $X = \{S\}$ and *σ-order-dense* if $X = \{S\}_\sigma$.

(10.6) PROPOSITION. *Let (X, C) be a Riesz space, S an ℓ-ideal in X, and suppose that $u \in C$. Then the following statements hold:*

(a) *$u \in \{S\}$ if and only if there exists a positive increasing net $\{u_\tau\}$ in S such that $u_\tau \uparrow u$.*

(b) *$u \in \{S\}_\sigma$ if and only if there exists a positive increasing sequence $\{u_n\}$ in S such that $u_n \uparrow u$.*

Proof. The proof of (b) is similar to that of (a) and will be omitted. For the proof of (a), it is clear that if $u_\tau \uparrow u$ with $u_\tau \in S$ then $u \in \{S\}$. Conversely, suppose that

$$B = \{u \in C : \text{there exists an increasing net } u_\tau \text{ in } S \cap C \text{ such that } u_\tau \uparrow u\}$$

and that
$$\hat{S} = B - B.$$

Then $S \subset \hat{S} \subset \{S\}$ and

$$B = \{u \in C : u = \sup \{w \in S : 0 \leqslant w \leqslant u\}\} \qquad (10.3)$$

because S is an ℓ-ideal in X. Notice that $\cup\{[0, u] : u \in B\} \subset B$; it then follows that
$$B = \{s^+ : s \in \hat{S}\}.$$

We complete the proof by showing that $\hat{S}$ is a normal subspace of X. If u_1, u_2 are in B, then $u_1 + u_2$ is an upper bound of the set

$$\{w \in S : 0 \leqslant w \leqslant u_1 + u_2\}.$$

On the other hand, if v is any upper bound of the set

$$\{w \in S : 0 \leqslant w \leqslant u_1 + u_2\},$$

then $v \geqslant w_1 + w_2$ for any $w_1, w_2 \in S$ with $0 \leqslant w_1 \leqslant u_1$, $0 \leqslant w_2 \leqslant u_2$. According to formula (10.3), there are

$$u_1 = \sup\{m \in S : 0 \leqslant m \leqslant u_1\} \quad \text{and} \quad u_2 = \sup\{n \in S : 0 \leqslant n \leqslant u_2\},$$

it follows that $u_1 + u_2 \leqslant v$, and hence that

$$u_1 + u_2 = \sup\{w \in S : 0 \leqslant w \leqslant u_1 + u_2\}.$$

This shows that $B+B \subset B$. Clearly $\lambda B \subseteq B$ for each non-negative λ and it follows that $\hat{S}$ is a vector subspace of X. Furthermore, $\hat{S}$ is an ℓ-ideal. If $y \in \hat{S}$ and if $x \in X$ is such that $|x| < |y|$, there exist $u_1, u_2 \in B$ such that $y = u_1 - u_2$. Then $u_1 + u_2 \in B$, and so $x^+ \in B$ by the construction of B and $0 \leqslant x^+ \leqslant u_1 + u_2$. Similarly, $x^- \in B$. Therefore $x = x^+ - x^- \in B - B = \hat{S}$; this shows that $\hat{S}$ is an ℓ-ideal. We now claim that $\hat{S}$ is a normal subspace of X. Suppose that $0 \leqslant u_\tau \uparrow u$ in X, where $u_\tau \in B$. Then

$$u = \sup u_\tau = \sup_\tau \sup\{v \in S : 0 \leqslant v \leqslant u_\tau\}$$
$$= \sup\{v \in S : 0 \leqslant v \leqslant u_\tau \quad \text{for some } \tau\}.$$

Note that for a fixed τ, the set $\{s \in S : 0 \leqslant s \leqslant u_\tau\}$ is directed upwards; therefore $u \in B$ and *a fortiori* $u \in \hat{S}$. This shows that $\hat{S}$ is a normal subspace of X, and the proof is complete.

(10.7) PROPOSITION.   *For any set A in a Riesz space X, the set A^d is always a normal subspace of X. Consequently $\{S\} \subseteq S^{dd}$ whenever $S \subseteq X$.*

Proof.   It is easy to verify the first assertion. To see the second, we note that S^{dd} is a normal subspace containing S; hence S^{dd} contains the normal subspace $\{S\}$ generated by S.

It should be noted that $\{S\}$ and S^{dd} are, in general, not equal, as shown by the following example: consider $\mathbf{R}^2$ equipped with the lexicographic ordering. Then $\mathbf{R}^2$ is a Riesz space which is not Archimedean since $0 \leqslant n(0, 1) \leqslant (1, 0)$ for all n; the only ℓ-ideals in $\mathbf{R}^2$ are $\{0\}$, the y-axis, and $\mathbf{R}^2$ itself; all of these ℓ-ideals are normal subspaces of $\mathbf{R}^2$. Now if we take S to be the y-axis, i.e.,

$$S = \{(0, y) : y \in \mathbf{R}\},$$

then $S = \{S\}$, $S^d = \{0\}$, and $S^{dd} = \mathbf{R}^2$; it then follows that $S \neq S^{dd}$.

(10.8) LEMMA.   *Let L_1, L_2 be subspaces of a Riesz space (X, C) such that $L_1 \cap L_2 = \{0\}$, and let L be the algebraic direct sum of L_1 and L_2. If L_1 and L_2 are ℓ-ideals in X, then L is the ordered direct sum of L_1 and L_2 (i.e. if $x_1 \in L_1$ and $x_2 \in L_2$, then $x_1 + x_2 \geqslant 0$ if and only if $x_1 \geqslant 0$ and $x_2 \geqslant 0$), denoted by $L = L_1 \oplus L_2$.*

Proof. Suppose that $x_i \in L_i$ $(i = 1, 2)$. It is clear that $x_1 + x_2 \geq 0$ whenever $x_i \geq 0$ $(i = 1, 2)$. Conversely, if $x_1 + x_2 \geq 0$ then

$$0 \leq x_1^- + x_2^- \leq x_1^+ + x_2^+, \qquad x_1^+, x_1^- \in L_1, \quad \text{and} \quad x_2^+, x_2^- \in L_2$$

because L_1 and L_2 are ℓ-ideals in X; in particular,

$$0 \leq x_1^- \leq x_1^+ + x_2^+.$$

By the Riesz decomposition property of X, there exist w_1, w_2 with $0 \leq w_i \leq x_i^+$ $(i = 1, 2)$ such that $x_1^- = w_1 + w_2$; therefore $w_i \in L_i$ $(i = 1, 2)$ since L_i are ℓ-ideals in X. It now follows from

$$w_2 = x_1^- - w_1 \in L_1$$

and from $L_1 \cap L_2 = \{0\}$ that $w_2 = 0$, and hence

$$x_1 = x_1^+ - x_1^- = x_1^+ - w_1 \geq 0.$$

Similarly we can show that $x_2 \geq 0$. Therefore L is the ordered direct sum of L_1 and L_2.

(10.9) PROPOSITION. *Let (X, C) be a Riesz space. Then the following statements are equivalent:*

(a) (X, C) *is Archimedean;*

(b) *for any ℓ-ideal A in X, if $A^{\mathrm{d}} = \{0\}$ then A is order-dense;*

(c) *for any ℓ-ideal A in X, $A \oplus A^{\mathrm{d}}$ is order-dense;*

(d) *for any normal subspace B of X, we have $B = B^{\mathrm{dd}}$;*

(e) *for any ℓ-ideal A in X, we have $\{A\} = A^{\mathrm{dd}}$.*

Proof. (a) $\Rightarrow$ (b): Let A be an ℓ-ideal in X such that $A^{\mathrm{d}} = \{0\}$. In view of proposition (10.6), we have to show, for each $u \in C$, that

$$u = \sup\{v \in A : 0 \leq v \leq u\}.$$

Let $A_u = \{v \in A : 0 \leq v \leq u\}$. Then u is certainly an upper bound of A_u. Let w be another upper bound of A_u such that $w \neq u$. We have to show that $u \leq w$. To do this, let $w' = u \wedge w$. Then w' is again an upper bound of A_u. We claim that $w' = u$ (so $u \leq w$). In fact, if $w' \neq u$, then let $z = u - w' > 0$. Since $A^{\mathrm{d}} = \{0\}$, $z \notin A^{\mathrm{d}}$ so there exists $a_0 \in A$ such that $z \wedge |a_0| \neq 0$. Let $z' = z \wedge |a_0|$. Since $z' = z \wedge |a_0| \leq |a_0| \in A$ and A is an ℓ-ideal, we have $z' \in A$; also $z' \leq z \leq u$, it follows that $z' \in A_u$. Moreover, since $z' + w' \leq z + w' = u$ and w' is an upper bound of A_u, we have $z' + A_u \subseteq A_u$ and *a fortiori* $2z' \in A_u$. Inductively we have $nz' \in A_u$, i.e. $nz' \leq u$ for all n; but $z' \neq 0$, contrary to (a). This proves the implication (a) $\Rightarrow$ (b).

(b) $\Rightarrow$ (c): Observe first that if A is an ℓ-ideal then so is the direct sum $A \oplus A^{\mathrm{d}}$. Further, $(A \oplus A^{\mathrm{d}})^{\mathrm{d}} = A^{\mathrm{d}} \cap A^{\mathrm{dd}} = \{0\}$; hence $A \oplus A^{\mathrm{d}}$ is order-dense by (b).

(c) $\Rightarrow$ (d): By proposition (10.7), $B = \{B\} \subseteq B^{\mathrm{dd}}$. Since both sets are positively generated, to see the opposite inclusion it is sufficient to show that $C \cap B^{\mathrm{dd}} \subseteq B$. Let $u \in C \cap B^{\mathrm{dd}}$. We have to show that $u \in B$. By (c), $B \oplus B^{\mathrm{d}}$ is order-dense; hence there exists a net $u_\tau \geqslant 0$ in $B \oplus B^{\mathrm{d}}$ such that $u_\tau \uparrow u$. For each τ, suppose $u_\tau = x_\tau + y_\tau$, where $0 \leqslant x_\tau \in B$ and $0 \leqslant y_\tau \in B^{\mathrm{d}}$. Notice that

$$0 \leqslant y_\tau \leqslant u_\tau \leqslant u \in B^{\mathrm{dd}},$$

hence $y_\tau \in B^{\mathrm{dd}}$. But we also have $y_\tau \in B^{\mathrm{d}}$; therefore $y_\tau = 0$, valid for all τ. Thus $x_\tau = u_\tau \uparrow u$ and $x_\tau \in B$; it follows that $u \in B$ because B is a normal subspace of X.

(d) $\Rightarrow$ (e): We start by observing that $A^{\mathrm{d}} = \{A\}^{\mathrm{d}}$. Next if $B = \{A\}$ then B is a normal subspace of X; hence, by (d), we have

$$\{A\} = B = B^{\mathrm{dd}} = \{A\}^{\mathrm{dd}} = A^{\mathrm{dd}}.$$

(e) $\Rightarrow$ (a): Suppose that X is non-Archimedean. Then there exist u, v in C such that $0 \lneqq nv \leqslant u$ for all n. Let $X_v = \bigcup n[-v, v]$. Then X_v is an ℓ-ideal in X. We shall show that $\{X_v \oplus X_v^{\mathrm{d}}\}$ is not $(X_v \oplus X^{\mathrm{d}})^{\mathrm{dd}}$. We verify this by showing that $u \notin \{X_v \oplus X^{\mathrm{d}}\}$ because of

$$(X_v \oplus X_v^{\mathrm{d}})^{\mathrm{dd}} = (0)^d = X.$$

Let
$$B_u = \{w \in X_v \oplus X_v^{\mathrm{d}} : 0 \leqslant w \leqslant u\}.$$

Then $(u-v)$ is an upper bound of B_u; in fact, let $w \in B_u$. Then w has a unique decomposition of the form $w = w_1 + w_2$ where $w_1 \in X_v$ and $w_2 \in X^{\mathrm{d}}$. Since X_v is an ℓ-ideal, we have that $w_1 + v \in X_v$, and hence that $w + v = (w_1 + v) + w_2 \in X_v \oplus X_v^{\mathrm{d}}$. Further, since $X_v \perp X^{\mathrm{d}}$,

$$(w_1 + v) \wedge w_2 = 0$$

so
$$w + v = (w_1 + v) + w_2 = (w_1 + v) \vee w_2 \leqslant u.$$

This shows that $(u-v)$ dominates every element w in B_u. Therefore $(u-v)$ is an upper bound of B_u and u is not the supremum of B_u. In view of the proof of proposition (10.6), we conclude that $u \notin \{X_v \oplus X_v^{\mathrm{d}}\}$.

If X is an order-complete Riesz space and A a subset of X, then A^{d} and A^{dd} are disjoint normal subspaces of X, and hence the algebraic direct sum of A^{d} and A^{dd} is the ordered direct sum, and $A^{\mathrm{d}} \oplus A^{\mathrm{dd}}$ is a subspace of X. In fact, $X = A^{\mathrm{d}} \oplus A^{\mathrm{dd}}$, as the following result, due to Riesz (1940) shows.

(10.10) PROPOSITION. *Let (X, C) be an order-complete Riesz space, and let A be a subset of X. Then X is the ordered direct sum of normal subspaces A^d and A^{dd} of X, that is, $X = A^d \oplus A^{dd}$.*

Proof. Given a positive element u in X, we show that $u = u_1 + u_2$, where $u_1 \in A^d$, $u_2 \in A^{dd}$, and $u_1 \geqslant 0$, $u_2 \geqslant 0$. Define u_1 by

$$u_1 = \sup\{u \wedge |x| : x \in A^d\}.$$

Since u is an upper bound of the set $\{u \wedge |x| : x \in A^d\}$, it follows from the order completeness of X that u_1 exists, and hence that $u_1 \in A^d$ because A^d is always a normal subspace of X. Notice that u_1 is the largest element in A^d which is majorized by u. It is clear that $0 \leqslant u_1 \leqslant u$. Let $u_2 = u - u_1$, then $u_2 \geqslant 0$. We further show that $u_2 \in A^{dd}$. To see this, let $x \in A^d$, and let $z = (u - u_1) \wedge |x|$. Since A^d is an ℓ-ideal, $z \in A^d$, thus $z + u_1 \in A^d$. Notice also that the element $z + u_1$ of A^d is dominated by u; hence $z + u_1 \leqslant u_1$ and $z \leqslant 0$. Since it is obvious that $z \geqslant 0$, we must have $z = 0$ and $z = u_2 \wedge |x| = 0$. This shows that $u_2 \in A^{dd}$, and hence that $u = u_1 + u_2 \in A^d \oplus A^{dd}$. Therefore $X = A^d \oplus A^{dd}$.

(10.11) COROLLARY. *Let (X, C) be an order-complete Riesz space. For any normal subspace B of X, we have $X = B \oplus B^d$.*

Proof. We remark that, since X is order-complete, the ordering in X must be Archimedean. Thus the result follows from propositions (10.9) and (10.10).

Let (X, C) be an order-complete Riesz space, and let B a normal subspace of X. Define P_B by

$$P_B(x) = \sup\{x^+ \wedge |y| : y \in B\} - \sup\{x^- \wedge |y| : y \in B\}.$$

According to proposition (10.10), $P_B(x) \in B$, and so P_B is a linear transformation of X onto B. This P_B is referred to as an ℓ-*projection* on B.

(10.12) THEOREM. *Let (E, C) be a weakly Riesz space. Then (E^b, C^*) is an order-complete Riesz space, and the following statements hold:*

(a) *for any $u \in C$ and $f \in E^b$, we have*

$$f^+(u) = \sup\{f(x) : 0 \leqslant x \leqslant u\},$$
$$f^-(u) = \sup\{f(x) : -u \leqslant x \leqslant 0\},$$
$$|f|(u) = \sup\{f(x) : -u \leqslant x \leqslant u\} = \sup\{|f(x)| : -u \leqslant x \leqslant u\},$$
$$|f(y)| \leqslant |f|(|y|) \quad \text{whenever } y \in E,$$

(b) *if $\{f_\tau : \tau \in D\}$ is a majorized increasing net in E^b, then*

$$h = \sup\{f_\tau : \tau \in D\}$$

exists in E^b, where $h(u) = \sup\{f_\tau(u) : \tau \in D\}$ for any u in C.

Proof. (a) We have shown in theorem (1.10) that (E^b, C^*) is a Riesz space and

$$f^+(u) = g(u) = \sup\{f(x) : 0 \leqslant x \leqslant u\} \qquad (u \in C).$$

Therefore, in view of $f^- = (-f)^+$, we obtain

$$\begin{aligned} f^-(u) = (-f)^+(u) &= \sup\{-f(y) : 0 \leqslant y \leqslant u\} \\ &= \sup\{f(-y) : -u \leqslant -y \leqslant 0\} \\ &= \sup\{f(x) : -u \leqslant x \leqslant 0\}. \end{aligned}$$

Since $|f| = 2f^+ - f$ then

$$\begin{aligned} |f|(u) = 2f^+(u) - f(u) &= 2\sup\{f(x) : 0 \leqslant x \leqslant u\} - f(u) \\ &= \sup\{f(2x - u) : 0 \leqslant x \leqslant u\} \\ &= \sup\{f(y) : -u \leqslant y \leqslant u\} \\ &= \sup\{|f(y)| : -u \leqslant y \leqslant u\}, \end{aligned}$$

in particular,

$$|f|(|y|) = \sup\{|f(w)| : -|y| \leqslant w \leqslant |y|\} \geqslant |f(y)|.$$

(b) Given $u \in C$, we define

$$\tilde{h}(u) = \sup\{f_\tau(u) : \tau \in D\}.$$

Then $\tilde{h}$ is positively homogeneous and additive on C because $f_\tau \uparrow$. Since C is generating, $\tilde{h}$ can be extended uniquely to a linear functional h on E. Notice that, since $\{f_\tau\}$ is majorized in E^b, $h \in E^b$. It is clear that $f_\tau \leqslant h$ for all $\tau \in D$. On the other hand, if $g \in E^b$ is such that $f_\tau \leqslant g$ for all $\tau \in D$, then $h(u) = \sup\{f_\tau(u) : \tau \in D\} \leqslant g(u)$ for all $u \in C$, and so $h = \sup\{f_\tau : \tau \in D\}$. Therefore (E^b, C^*) is an order-complete Riesz space.

Remark. If f, g are in E^b we have, by making use of the preceding result, that

(a) $\begin{aligned}[t] (f \vee g)(u) &= \sup\{f(x) + g(u - x) : 0 \leqslant x \leqslant u\} \\ &= \sup\{f(v) + g(w) : v, w \geqslant 0 \quad \text{and} \quad u = v + w\}; \end{aligned}$

(b) $\begin{aligned}[t] (f \wedge g)(u) &= \inf\{f(x) + g(u - x) : 0 \leqslant x \leqslant u\} \\ &= \inf\{f(v) + g(w) : v, w \geqslant 0 \quad \text{and} \quad u = v + w\} \quad (u \in C); \end{aligned}$

(c) for any $u \in C$, $[-u, u]^\pi = \{f \in E^b : |f|(u) \leqslant 1\}$, where $[-u, u]^\pi$ is the polar of the set $[-u, u]$ taken in E^b.

Recall that a semi-norm p on an ordered vector space (E, C) is a Riesz semi-norm if it satisfies the following two conditions:

(i) p is absolute-monotone, i.e. $-u \leqslant x \leqslant u$ in $E \Rightarrow p(x) \leqslant p(u)$;

(ii) for each $x \in E$ with $p(x) < 1$ there exists $u \in E$ with $p(u) < 1$ such that $-u \leqslant x \leqslant u$.

In view of lemma (9.16), if (X, C) is a Riesz space, then a semi-norm p on X is a Riesz semi-norm if and only if it follows from $|x| \leqslant |y|$ with x, y in X that $p(x) \leqslant p(y)$.

(10.13) THEOREM (Luxemburg–Zaanen). *Let (X, C) be a Riesz space, p a monotone semi-norm on X, and suppose that Y is a Riesz subspace of X. If f is a positive linear functional on Y which is dominated by p on Y, then there exists a positive and linear extension ϕ of f such that $|\phi(x)| \leqslant p(|x|)$ for any $x \in X$. Furthermore, if p is a Riesz semi-norm then ϕ is dominated by p.*

Proof. Suppose that

$$q(x) = p(x^+) \qquad (x \in X).$$

It is easily seen that q is sublinear on X and that

$$f(y) \leqslant f(y^+) \leqslant p(y^+) = q(y) \quad \text{for all } y \in Y.$$

In view of the Hahn–Banach extension theorem, there exists $\phi \in X^*$ such that $\phi(y) = f(y)$ for all $y \in Y$ and $\phi(x) \leqslant q(x)$ for all $x \in X$. For any $u \in C$, we have

$$\phi(-u) \leqslant q(-u) = p((-u)^+) = p(0) = 0,$$

and so $\phi \in C^*$. On the other hand, by theorem (10.12)(a), we have

$$|\phi(x)| \leqslant |\phi|(|x|) = \phi(|x|) \leqslant q(|x|) = p(|x|) \quad \text{for all } x \in X.$$

In particular, if p is a Riesz semi-norm then

$$|\phi(x)| \leqslant p(|x|) = p(x),$$

and so ϕ is dominated by p. This completes the proof.

Before giving a dual result of theorem (10.12)(a), we need the following lemma.

(10.14) LEMMA. *Let (X, C) be a Riesz space, and let $f \in C^*$. For any $u \in C$, there exists $g \in C^*$ such that $0 \leqslant g \leqslant f$, $g(u) = f(u)$, and $g(x) = 0$, whenever $x \perp u$.*

Proof. For any $w \in C$, define

$$h(w) = \sup_n f(w \wedge nu).$$

Then h is positively homogeneous on C, $0 \leqslant h(w) \leqslant f(w)$ for all $w \in C$, and $h(u) = f(u)$; further, h is additive on C. In fact, if w_1 and w_2 are in C, by proposition (10.1)(b), we have

$$(w_1 + w_2) \wedge nu \leqslant w_1 \wedge nu + w_2 \wedge nu$$

and so

$$h(w_1 + w_2) \leqslant h(w_1) + h(w_2).$$

On the other hand, it is clear that

$$w_1 \wedge nu + w_2 \wedge mu \leqslant (w_1 + w_2) \wedge (n + m)u,$$

and thus

$$h(w_1) + h(w_2) \leqslant h(w_1 + w_2).$$

There exists a linear functional g on X such that $g(w) = h(w)$ for all $w \in C$; in particular,

$$g(u) = h(u) = f(u), \qquad 0 \leqslant g(w) = h(w) \leqslant f(w) \quad \text{for all } w \in C.$$

Finally, if $|x| \wedge u = 0$, then $|x| \wedge nu = 0$ for all n, consequently

$$0 \leqslant |g(x)| \leqslant g(|x|) = \sup_n f(|x| \wedge nu) = 0.$$

Now we easily deduce a dual result of theorem (10.12)(a) as follows.

(10.15) PROPOSITION. *Let (X, C) be a Riesz space. For any $f \in C^*$ and $x \in X$, the following equalities hold:*
(a) $f(x^+) = \sup\{g(x) : 0 \leqslant g \leqslant f\}$;
(b) $f(x^-) = \sup\{h(x) : -f \leqslant h \leqslant 0\}$;
(c) $f(|x|) = \sup\{g(x) : -f \leqslant g \leqslant f\} = \sup\{|g(x)| : -f \leqslant g \leqslant f\}$.

Proof. It is clear that (b) and (c) follow from (a), and hence we only have to show the assertion (a). For any $g \in E^b$ with $0 \leqslant g \leqslant f$, the following inequalities hold:

$$g(x) \leqslant g(x^+) \leqslant f(x^+),$$

hence $\sup\{g(x) : 0 \leqslant g \leqslant f\} \leqslant f(x^+)$. Apply lemma (10.14) to obtain $h \in C^*$ such that $0 \leqslant h \leqslant f$, $h(x^+) = f(x^+)$, and $h(y) = 0$, whenever $y \perp x^+$. Therefore

$$f(x^+) = h(x^+) = h(x^+) - h(x^-) = h(x),$$

consequently $f(x^+) = \sup\{g(x):0 \leqslant g \leqslant f\}$.

This completes the proof of this proposition.

Remark. For any $f \in C^*$, $[-f,f]^0 = \{x \in X : f(|x|) \leqslant 1\}$, where, of course, the polar is taken in X.

The following result is a consequence of theorem (10.12)(a) and proposition (10.15).

(10.16) PROPOSITION. *Let (X, C) be a Riesz space and let Y be a Riesz subspace of (X^b, C^*). Then the following statements hold:*

(a) *if A is a solid subset of X then the polar $A^\pi(Y)$ of A, taken in Y, is a solid subset of Y;*

(b) *if B is a solid subset of X^b then the polar B^π of B, taken in X, is a solid subset of X.*

Proof. Let $g \in A^\pi(Y)$, and let $f \in Y$ be such that $|f| \leqslant |g|$. For any $a \in A$, we have, by theorem (10.12)(a), that

$$f(a) \leqslant |f|(|a|) \leqslant |g|(|a|) = \sup\{g(x) : |x| \leqslant |a|\}.$$

Since A is a solid subset of X and since $g \in A^\pi(Y)$, it follows that $|g|(|a|) \leqslant 1$, and hence that $f(a) \leqslant 1$. Therefore $A^\pi(Y)$ is a solid subset of Y. This proves the assertion (a). The proof of (b) is similar, by making use of proposition (10.15).

Let (X, C) be a Riesz space with the order-bound dual X^b, and let ϕ be in X^b. ϕ is called a *normal integral* (or *order-continuous*) on X if for any net $\{u_\tau\}$ with $u_\tau \downarrow 0$, then $\inf |\phi(u_\tau)| = 0$; ϕ is called an *integral* (or *order-σ-continuous*) if, for any sequence $\{u_n\}$ with $u_n \downarrow 0$, then $\inf |\phi(u_n)| = 0$. The set of all normal integrals on X is denoted by X_n^b and the set of all integrals on X is denoted by X_c^b. It is clear that $X_n^b \subset X_c^b$.

(10.17) PROPOSITION. *Let (X, C) be a Riesz space. Then X_n^b and X_c^b are normal subspaces of X^b.*

Proof. It is clear that X_c^b is a vector subspace of X^b, and that if $0 \leqslant g \leqslant h \in X_c^b$ then $g \in X_c^b$. In order to show that X_c^b is an ℓ-ideal, it suffices to verify that if $f \in X_c^b$ then $f^+ \in X_c^b$. Suppose $u_n \downarrow 0$. For any w in X with $0 \leqslant w \leqslant u_1$, we have, by proposition (10.1)(a), that

$$0 \leqslant w - w \wedge u_n \leqslant u_1 - u_n \quad \text{and} \quad w \wedge u_n \downarrow 0.$$

It follows from $f(w-w \wedge u_n) \leqslant f^+(w-w \wedge u_n) \leqslant f^+(u_1-u_n)$ that

$$0 \leqslant f^+(u_n) \leqslant f^+(u_1)-f(w)+f(w \wedge u_n) \quad \text{for all } n,$$

and hence that

$$0 \leqslant \inf f^+(u_n) \leqslant f^+(u_1)-f(w) \quad \text{for all } w \in [0, u_1].$$

By theorem (10.12)(a), $\inf f^+(u_n) = 0$, and so $f^+ \in X_c^b$.

Now suppose that $\{f_\tau\}$ is a positive increasing net in X_c^b such that $f_\tau \uparrow f$ in X^b, and that $u_n \downarrow 0$ in X. By theorem (10.12)(b),

$$f(u_1) = \sup f_\tau(u_1).$$

For any $\varepsilon > 0$, there exists τ_0 such that $0 \leqslant f(u_1)-f_{\tau_0}(u_1) < \varepsilon$. Since $u_n \downarrow 0$ and since $f-f_{\tau_0} \geqslant 0$, it follows that

$$0 \leqslant f(u_n)-f_{\tau_0}(u_n) \leqslant f(u_1)-f_{\tau_0}(u_1) < \varepsilon \quad \text{for all } n,$$

and hence that $\lim_n[f(u_n)-f_{\tau_0}(u_n)] \leqslant \varepsilon$. Notice that $0 \leqslant f_{\tau_0} \in X_c^b$ and that

$$\lim_n f_{\tau_0}(u_n) = \inf_n f_{\tau_0}(u_n) = 0.$$

Therefore $\inf_n f(u_n) = \lim_n f(u_n) \leqslant \varepsilon$. Since ε is arbitrary, this implies that $f \in X_c^b$, and so X_c^b is a normal subspace of X^b. This proves the result for X_c^b, and the proof for X_n^b is similar.

Let (X, C) be a Riesz space. For any $f \in X^b$, we define

$$N_f = \{x \in X : |f|\,(|x|) = 0\}.$$

It is easily seen that N_f is the largest ℓ-ideal in X contained in the kernel of f, and that if f is a normal integral then N_f is a normal subspace of X. Furthermore, we also have that $N_f = N_{|f|} = N_{f^+} \cap N_{f^-}$, and that $N_f = \mathrm{sk}(f^{-1}(0))$.

The following two results, which will be needed later, are of some interest in themselves.

(10.18) PROPOSITION. *Let* (X, C) *be a Riesz space,* $h \in X^b$, *and let* $\phi \in X_c^b$. *If* $h \perp \phi$ *then* $N_h^d \subset N_\phi$.

Proof. Without loss of generality one can assume that h and ϕ are positive. Notice that the statement that $N_h^d \subset N_\phi$ is equivalent to the statement that if $0 \leqslant w \in N_h^d$ then $w \in N_\phi$ because N_h^d and N_ϕ are ℓ-ideals in X. We now suppose $0 \leqslant w \in N_h^d$. Since $\phi \wedge h = 0$ and since

$$(\phi \wedge h)(w) = \inf\{\phi(x)+h(y) : x, y \in C,\ x+y = w\},$$

there exist positive sequences $\{x_n\}$ and $\{y_n\}$ with $x_n + y_n = w$ such that

$$\phi(x_n) + h(y_n) \leqslant 1/2^n \quad \text{for all } n = 1, 2, \ldots .$$

For any $\varepsilon > 0$, if we can find positive sequences $\{y_n'\}$, $\{y_n''\}$, and $\{z_n\}$ with the following properties:

(a) $0 \leqslant y_n' \leqslant z_n \downarrow 0, \qquad y_n = y_n' + y_n''$,

(b) $\phi(y_n'') < \varepsilon$,

then $\phi(w) = 0$, and so $w \in N_\phi$; in fact, since $x_n = w - y_n = w - y_n' - y_n''$, we then have

$$\phi(w - y_n') = \phi(x_n + y_n'') \leqslant 2^{-n} + \varepsilon \quad \text{for all } n = 1, 2, \ldots .$$

Note also that $w \geqslant w - y_n' \geqslant w - z_n \uparrow w$; it follows from $0 \leqslant \phi \in X_c^b$ that $\phi(w - y_n') \to \phi(w)$ as $n \to \infty$, and hence that $\phi(w) \leqslant \varepsilon$; consequently $\phi(w) = 0$.

We complete the proof by showing the existence of positive sequences $\{y_n'\}$, $\{y_n''\}$, and $\{z_n\}$ satisfying (a) and (b). For each $n \geqslant 1$, let

$$y_{n,m} = \sup\{y_n, y_{n+1}, \ldots, y_m\} \qquad (m \geqslant n).$$

Then $y_n \leqslant y_{n,m} \leqslant w$ for all n, $y_{n,m}$ are in N_h^d for all n, m, and $y_{k,m} \geqslant y_{j,m}, y_{n,k} \leqslant y_{n,j}$ whenever $k < j$, so that $\phi(y_{n,m}) \uparrow_m \leqslant \phi(w)$. It follows that $\lim_{m \to \infty} \phi(y_{n,m})$ exists, and hence that there exists a sequence $\{m_n\}$ of natural numbers with $m_n \uparrow$ such that

$$\phi(y_{n,k}) - \phi(y_{n,m_n}) < \varepsilon/2^n \quad \text{for all } k \geqslant m_n.$$

Let

$$z_n = \inf\{y_{1,m_1}, y_{2,m_2}, \ldots, y_{n,m_n}\}.$$

Then $z_n \downarrow$, $z_n \leqslant y_{n,m_n}$, and so $z_n \in N_h^d$; furthermore $z_n \downarrow 0$. In fact, since

$$z_n \leqslant y_{n,m_n} \leqslant y_n + y_{n+1} + \ldots + y_{m_n},$$

we have that

$$0 \leqslant h(z_n) \leqslant 2^{-n} + 2^{-(n+1)} + \ldots + 2^{-m_n} \to 0 \quad (\text{as } n \to \infty).$$

Hence if $v \in C$ is a lower bound of the set $\{z_n\}$ then $h(v) = 0$, which is to say that $v \in N_h$. Notice that $0 \leqslant v \leqslant z_n \in N_h^d$ and that N_h^d is an ℓ-ideal, hence $v \in N_h^d$. Consequently $v = 0$.

We now define

$$y_n' = y_n \wedge z_n \quad \text{and} \quad y_n'' = y_n - y_n'.$$

Then $0 \leqslant y_n' \leqslant z_n \downarrow 0$, $y_n'' \geqslant 0$, and $y_n = y_n' + y_n''$; it remains to show that $\phi(y_n'') < \varepsilon$. For each n, let

$$v_n = \sum_{i=1}^{n-1} (y_{i,m_n} - y_{i,m_i}).$$

Then

$$\phi(v_n) = \sum_{i=1}^{n-1} [\phi(y_{i,m_n}) - \phi(y_{i,m_i})] < \sum_{i=1}^{n-1} 2^{-i}\varepsilon < \varepsilon \quad \text{for all } n.$$

Since $y_{k,m} \geqslant y_{j,m}$ and $y_{n,k} \leqslant y_{n,j}$ whenever $k \leqslant j$, it follows that $v_n \geqslant 0$ and

$$y_{n,m_n} - v_n \leqslant y_{i,m_i} \quad \text{for all } i = 1, 2, \ldots, n,$$

and hence from the definition of z_n that

$$y_{n,m_n} - v_n \leqslant z_n \quad \text{for all } n.$$

We observe that $y_n \leqslant y_{n,m_n}$, thus

$$y_n - v_n \leqslant z_n \quad \text{for all } n.$$

On the other hand, since

$$y_n'' = y_n - y_n' = y_n - y_n \wedge z_n = \sup\{0, y_n - z_n\} \leqslant v_n,$$

since $\phi \geqslant 0$, and since $\phi(v_n) < \varepsilon$ for all n, then

$$\phi(y_n'') \leqslant \phi(v_n) < \varepsilon \quad \text{for all } n,$$

and so $\{y_n'\}$, $\{y_n''\}$, and $\{z_n\}$ are the required positive sequences.

A partial converse of the preceding result is given below.

(10.19) COROLLARY.   *Let (X, C) be an Archimedean Riesz space, and let h, ϕ be in X_n^b. Then the following statements are equivalent:*
  (a) $h \perp \phi$;
  (b) $N_h^d \subset N_\phi$;
  (c) $N_\phi^d \subset N_h$;
  (d) $N_h^d \perp N_\phi^d$.

Proof.   Since X is Archimedean, proposition (10.9) gives $\{A\} = A^{dd}$ for any ℓ-ideal A in X; in particular, $N_h = \{N_h\} = (N_h)^{dd}$ for every $h \in X_n^b$. Therefore (b), (c), and (d) are equivalent. On the other hand, the implication (a) $\Rightarrow$ (b) follows from proposition (10.18). Conversely if $N_h^d \subset N_\phi$, then $|\phi| \wedge |h| \in X_n^b$ and $(|\phi| \wedge |h|)(x) = 0$ for all $x \in N_h \oplus N_h^d$. Since X is Archimedean, it follows from proposition (10.9) that $N_h \oplus N_h^d$ is order-dense, and hence that $|\phi| \wedge |h| = 0$ on X; i.e. $\phi \perp h$.

(10.20) PROPOSITION.   *Let (X, C) be an Archimedean Riesz space, $0 \leqslant h \in X_n^b$, and let $u \in C$ be such that $h(u) > 0$. Then there exists $w \in X$ with $0 \leqslant w \leqslant u$ such that $h(w) > 0$, and*

$$\phi(w) = 0 \quad \text{for all } \phi \in X_c^b \text{ with } \phi \perp h.$$

Proof. Since (X, C) is Archimedean, it follows from proposition (10.9) that $N_h \oplus N_h^d$ is order-dense, and hence from proposition (10.6) that
$$u = \sup\{s+w : 0 \leqslant s \in N_h, 0 \leqslant w \in N_h^d, s+w \leqslant u\}.$$

Since $0 \leqslant h \in X_n^b$, we have
$$h(u) = \sup\{h(s)+h(w) : 0 \leqslant s \in N_h, 0 \leqslant w \in N_h^d, s+w \leqslant u\}$$
$$= \sup\{h(w) : 0 \leqslant w \leqslant u, w \in N_h^d\},$$

and so, by making use of the fact that $h(u) > 0$, there exists $w \in N_h^d$ with $0 \leqslant w \leqslant u$ such that $h(w) > 0$. If $\phi \in X_c^b$ is such that $\phi \perp h$, then, by proposition (10.18), $N_h^d \subset N_\phi$, and so $w \in N_\phi$, because $w \in N_h^d$. We conclude from $|\phi(w)| \leqslant |\phi|(|w|)$ that $\phi(w) = 0$. This completes the proof.

(10.21) COROLLARY. *Let (X, C) be an Archimedean Riesz space, and let B be a normal subspace of X_n^b. If $0 \leqslant f \in X_n^b \backslash B$, then there exists $w \in C$ such that*
$$f(w) > 0 \quad and \quad w \in B^0,$$

where B^0 is the polar of B with respect to the duality $\langle X, X_n^b \rangle$.

Proof. Set $B^d = \{h \in X_n^b : h \perp B\}$. Since X_n^b is order-complete, it follows from corollary (10.11) that $X_n^b = B \oplus B^d$, and hence that f has the unique decomposition $f = g+h$, where $0 \leqslant g \in B$ and $0 \leqslant h \in B^d$. Note that $h \neq 0$ because $f \notin B$; hence there exists $u \in C$ such that $h(u) > 0$. Since $h \perp B$, and since $X_n^b \subset X_c^b$, we conclude from proposition (10.20) that there exists $w \in X$ with $0 \leqslant w \leqslant u$ such that $h(w) > 0$ and $w \in B^0$, and hence that $f(w) = g(w)+h(w) > 0$. This completes the proof.

Let A be a subset of a Riesz space (X, C). A_n^π denotes the polar of A taken in X_n^b, i.e.
$$A_n^\pi = \{f \in X_n^b : f(a) \leqslant 1 \; \forall \, a \in A\}.$$

A_c^π denotes the polar of A taken in X_c^b, i.e.
$$A_c^\pi = \{f \in X_c^b : f(a) \leqslant 1 \; \forall \, a \in A\}.$$

If A is a subspace of X, then
$$A_n^\pi = \{f \in X_n^b : f(a) = 0 \; \forall \, a \in A\}$$
and
$$A_c^\pi = \{f \in X_c^b : f(a) = 0 \; \forall \, a \in A\}.$$

(10.22) COROLLARY. *Let (X, C) be an Archimedean Riesz space. If B is an ℓ-ideal in X_n^b, then $\{B\} = (B^0)_n^\pi$, where*

$$B^0 = \{x \in X : f(x) = 0 \; \forall \, f \in B\}.$$

Proof. We start with the observation that $(B^0)_n^\pi$ is a normal subspace of X_n^b such that $B \subset (B^0)_n^\pi$; it then follows that $\{B\} \subset (B^0)_n^\pi$. Note that $B^0 = \{B\}^0$. If $0 < f \in X_n^b \backslash \{B\}$, then, by corollary (10.21), there exists $w \in C$ such that
$$f(w) > 0 \quad \text{and} \quad w \in \{B\}^0;$$

hence $f \notin (B^0)_n^\pi$, consequently $\{B\} = (B^0)_n^\pi$. This completes the proof.

(10.23) COROLLARY. *Let (X, C) be an Archimedean Riesz space, and let A, B be ℓ-ideals in X_n^b. Then the following statements hold:*
   (a) $\{A\} \subset \{B\}$ *if and only if* $A^0 \supset B^0$;
   (b) $\{A\} = \{B\}$ *if and only if* $A^0 = B^0$.

Proof. The conclusion (b) is an immediate consequence of the assertion (a); and the assertion (a) follows from corollary (10.22). This completes the proof.

Let X, Y be Riesz spaces, and let T be a linear operator of X into Y. T is called a *lattice homomorphism* (or briefly an *ℓ-homomorphism*) if it preserves the lattice operations; and the ℓ-homomorphism is called an *ℓ-isomorphism* if it is injective. It is clear that each ℓ-homomorphism is positive.

We list some elementary properties of ℓ-homomorphisms as follows.

(10.24) PROPOSITION. *Let X, Y be Riesz spaces, and let T be an ℓ-homomorphism of X into Y. Then the following statements hold:*
   (a) *the kernel of T, written $k(T)$, is an ℓ-ideal in X;*
   (b) $T(x) \geqslant 0$ *if and only if there exists u in $C \cap k(T)$ such that $x + u \geqslant 0$;*
   (c) $T(x) \leqslant T(y)$ *if and only if there exists $z \in X$ such that $x \vee y \leqslant z$ and $T(y) = T(z)$;*
   (d) $|T(x)| \leqslant |T(y)|$ *if and only if there exists $w \in k(T)$ such that $|w| \leqslant |x|$ and $|x - w| \leqslant |y|$;*
   (e) *if A is a solid set in X then $T(A)$ is a solid set in $T(X)$;*
   (f) *if B is a solid set in $T(X)$ then so is $T^{-1}(B)$ in X.*

Proof. (a) Straightforward.

(b) The sufficiency is obvious. To prove its necessity, we note that

$$0 \leqslant T(x) = (T(x))^+ = T(x^+)$$

and that $x^- = x^+ - x \in C \cap k(T)$. Then $u = x^-$ is the required element.

(c) The sufficiency follows from $T(x) \leqslant T(z) = T(y)$. For the necessity, we observe that $T(y-x) \geqslant 0$, there exists, by (b), $u \in C \cap k(T)$ such that $y-x+u \geqslant 0$. Let $z = u+y$. Then $x \vee y \leqslant z$ and $T(y) = T(z)$.

(d) (i) *Necessity.* We start with the observation that there exists $v \in C \cap k(T)$ such that

$$|x| \leqslant |y|+v \quad \text{and} \quad 0 \leqslant v \leqslant |x|.$$

In fact, since $T(|x|) = |T(x)| \leqslant |T(y)| = T(|y|)$, there exists, by (b), $u \in C \cap k(T)$ such that $u+|y|-|x| \geqslant 0$, so $|x|-|y| \leqslant u$ and hence $|x|-|y| \leqslant u \wedge |x|$. Taking $v = u \wedge |x|$, then $|x|-|y| \leqslant v$, $0 \leqslant v \leqslant |x|$, and $v \in C \cap k(T)$ because $0 \leqslant v \leqslant u$.

On the other hand, since $0 \leqslant v \leqslant |x| = x^+ + x^-$, by the Riesz decomposition property of X there exist $v_1, v_2 \in X$ with $0 \leqslant v_1 \leqslant x^+$, $0 \leqslant v_2 \leqslant x^-$ such that $v = v_1 + v_2$. Obviously $v_1 \wedge v_2 = 0$. Set

$$w = v_1 - v_2.$$

We obtain, by virtue of proposition (10.3)(h),

$$|w| = |v_1 - v_2| = |v_1| + |v_2| = v_1 + v_2 = v,$$

so $w \in k(T)$, $|w| \leqslant |x|$. Similarly, since $(x^+ - v_1) \wedge (x^- - v_2) = 0$, we also have

$$|x-w| = |x^+ - x^- - v_1 + v_2| = (x^+ - v_1) + (x^- - v_2) = |x| - v,$$

which implies that $|x-w| \leqslant |y|$.

(ii) *Sufficiency.* Since $w \in k(T)$ and since

$$\big||x-w| - |x|\big| \leqslant |x-w-x| = |w|,$$

it follows that $|x-w| - |x| \in k(T)$. The sufficiency now follows from the following computation.

$$|T(x)| = T(|x|) = T(|x-w|) \leqslant T(|y|) = |T(y)|.$$

(e) Let $z \in T(X)$ be such that $|z| \leqslant |T(a)|$ for some $a \in A$. There exists $x \in X$ such that $z = T(x)$. By (d), there exists $w \in k(T)$ such that $|w| \leqslant |x|$ and $|x-w| \leqslant |a|$. Since A is solid, it follows that $x-w \in A$ and

10

hence that $T(x-w) \in T(A)$. We conclude from $T(x) = T(x-w)$ that $T(x) \in T(A)$. Therefore $T(A)$ is a solid subset of $T(X)$.

(f) Let z be in $T^{-1}(B)$, and let $x \in X$ be such that $|x| \leqslant |z|$. Then $|T(x)| = T(|x|) \leqslant T(|z|) = |T(z)|$, so $T(x) \in B$ because B is a solid subset of $T(X)$, and thus $x \in T^{-1}(B)$. Therefore $T^{-1}(B)$ is a solid subset of X. This completes the proof.

Let J be an ℓ-ideal in a Riesz space (X, C), and let $x \to [x]$ be the canonical mapping of X onto X/J defined by $[x] = x+J$. If

$$[C]_J = \{[x] \in X/J: \text{there exists } j \in J \text{ such that } x+j \in C\},$$

then $(X/J, [C]_J)$ is a Riesz space, and the canonical mapping $x \to [x]$ is an ℓ-homomorphism of X onto X/J.

If $\{(X_\alpha, C_\alpha): \alpha \in \Gamma\}$ is a family of Riesz spaces, then the product space $X = \prod_{\alpha \in \Gamma} X_\alpha$ is a Riesz space under the ordering $C = \prod_{\alpha \in \Gamma} C_\alpha$, and the algebraic direct sum $\bigoplus_{\alpha \in \Gamma} X_\alpha$ is an ℓ-ideal in (X, C). Moreover, each projection $\pi_\alpha: \prod_{\alpha \in \Gamma} X_\alpha \to X_\alpha$ is an ℓ-homomorphism; also each injection $j_\alpha: X_\alpha \to \bigoplus_{\alpha \in \Gamma} X_\alpha$ is an ℓ-isomorphism. Observe that the properties of order-completeness as well as of σ-order-completeness are preserved in the formation of products and algebraic direct sums.

Let (X, C) and (Y, K) be Riesz spaces. We say that (Y, K) is an *order-completion* of (X, C) if:

(a) (Y, K) is order-complete;

(b) there exists an ℓ-isomorphism, say $x \to \hat{x}$, of (X, C) into (Y, K);

(c) for any $y \in Y$, we have

$$y = \sup\{\hat{a}: a \in X, \hat{a} \leqslant y\} = \inf\{\hat{b}: b \in X, y \leqslant \hat{b}\}.$$

It should be noted that the condition (c) in the definition of order-completion can be replaced by the following condition:

(d) for any $y \in Y$ with $y > 0$, there exist a, b in X such that $0 < \hat{a} \leqslant y \leqslant \hat{b}$.

For a proof, see Luxemburg and Zaaren (1971).

For the sake of convenience, we shall identify X with the Riesz subspace $\{\hat{a}: a \in X\}$ of Y. The following result is concerned with the existence of the order-completion of a Riesz space. The construction of an order-completion of an Archimedean Riesz space is a straightforward generalization of the Dedekind procedure for completing the rational number system, therefore we leave the proof which can be found in Peressin (1967)) to the reader.

(10.25) THEOREM (Nakano). *A Riesz space* (X, C) *has an order-completion if and only if* X *is Archimedean. Furthermore any two order-completions of* X *are ℓ-isomorphic.*

As we shall see in Chapter 13, the order-completion of a locally convex Riesz space may be identified with the topological dual of some locally convex Riesz space under certain additional assumptions.

11

TOPOLOGICAL RIESZ SPACES

By a *topological Riesz space* (or *topological vector lattice*) we mean a Riesz space (X, C) equipped with a vector topology $\mathscr{P}$ which admits a neighbourhood-base at 0 consisting of solid sets in X. A Riesz space equipped with a locally solid topology is referred to as a *locally convex Riesz space* (or *locally convex vector lattice*). Therefore, by making use of theorem (6.3), $(X, C, \mathscr{P})$ is a locally convex Riesz space if and only if it is both locally o-convex and locally decomposable.

(11.1) PROPOSITION. *Let (X, C) be a Riesz space, and let $(X, \mathscr{P})$ be a topological vector space. Then the following statements are equivalent:*

(a) *$(X, C, \mathscr{P})$ is a topological Riesz space;*

(b) *the mapping $(x, y) \to x \vee y$ is uniformly continuous on $X \times X$;*

(c) *the mapping $x \to x^+$ is uniformly continuous on X;*

(d) *$(X, C, \mathscr{P})$ is locally full and C gives an open decomposition on $(X, \mathscr{P})$;*

(e) *$(X, C, \mathscr{P})$ is locally full and the mapping $x \to x^+$ is continuous at 0;*

(f) *for any two nets $\{x_\alpha : \alpha \in D\}$ and $\{y_\alpha : \alpha \in D\}$ in X, if $|x_\alpha| \leqslant |y_\alpha|$ for all $\alpha \in D$ and if y_α converges to 0 with respect to $\mathscr{P}$, then x_α converges to 0 with respect to $\mathscr{P}$.*

Proof. (a) $\Rightarrow$ (b): Let V be a solid $\mathscr{P}$-neighbourhood of 0, and let W be a solid $\mathscr{P}$-neighbourhood of 0 such that $W + W \subset V$. For any elements $x, y, w_1,$ and w_2 in X,

$$w_1 \wedge w_2 + x \wedge y \leqslant (x + w_1) \vee (y + w_2) \leqslant w_1 \vee w_2 + x \vee y,$$
$$-(|w_1| + |w_2|) \leqslant w_1 \wedge w_2 \leqslant w_1 \vee w_2 \leqslant |w_1| + |w_2|,$$

and so

$$-(|w_1| + |w_2|) \leqslant (x + w_1) \vee (y + w_2) - x \vee y \leqslant |w_1| + |w_2|.$$

Therefore, if $w_1, w_2 \in W$ then $(x + w_1) \vee (y + w_2) - x \vee y \in V$; this implies that the mapping $(x, y) \to x \vee y$ is uniformly continuous on $X \times X$.

(b) $\Rightarrow$ (c): Obvious.

(c) $\Rightarrow$ (d): Let W be a $\mathscr{P}$-neighbourhood of 0. By the uniform continuity of the mapping $x \to x^+$, there exists a $\mathscr{P}$-neighbourhood V of

0 such that $x^+ - y^+ \in W$ whenever $x - y \in V$. We now claim that $(V - C) \cap C \subset W$, and then it would follow from theorem (5.1) that $(X, C, \mathscr{P})$ is locally full. Suppose that $0 \leqslant w \leqslant u$, where $u \in V$ and $w \in X$. Then $w - (w - u) = u \in V$, and so $w = w^+ = w^+ - (w - u)^+ \in W$. For the proof of the open decomposition property, we first notice that the mapping $x \to x^+$ is continuous at 0. Let W be any $\mathscr{P}$-neighbourhood of 0; then there exists a $\mathscr{P}$-neighbourhood V of 0 such that $x^+, x^- \in W$ whenever $x \in V$, so $V \subset W \cap C - W \cap C$, and thus C gives an open decomposition on $(X, \mathscr{P})$.

(d) $\Rightarrow$ (e): Let W be a circled, order-convex $\mathscr{P}$-neighbourhood of 0, and let $V = W \cap C - W \cap C$. Then V is a $\mathscr{P}$-neighbourhood. For any $x \in V$, there exist w_1, w_2 in $W \cap C$ such that $x = w_1 - w_2$, so $0 \leqslant x^+ \leqslant w_1$, $0 \leqslant x^- \leqslant w_2$; it then follows from the order-convexity of W that $x^+ \in W$, and hence that the mapping $x \to x^+$ is continuous at 0.

(e) $\Rightarrow$ (a): For any $\mathscr{P}$-neighbourhood W of 0, there exists a circled order-convex $\mathscr{P}$-neighbourhood U of 0 such that $U \subset W$. Since the mapping $x \to x^+$ is continuous at 0, then the mapping $x \to |x|$ is also continuous at 0, so there exists a $\mathscr{P}$-neighbourhood V of 0 such that $|x| \in U$ whenever $x \in V$, and so $V \subset \text{sk}(W)$; this shows that the solid kernel $\text{sk}(W)$ of W is a $\mathscr{P}$-neighbourhood, consequently $(X, C, \mathscr{P})$ is a topological Riesz space.

Therefore statements (a)–(e) are mutually equivalent. Finally we show that these statements are equivalent to (f). The implication (a) $\Rightarrow$ (f) is easy to verify. On the other hand, suppose (f) holds. Then, in view of theorem (5.1), $\mathscr{P}$ must be locally full; further, the map $x \to x^+$ is clearly continuous. This shows that (f) $\Rightarrow$ (a); consequently statements (a)–(f) are equivalent.

If $(X, C, \mathscr{P})$ is an order-complete topological Riesz space and if B is a normal subspace of X then, by the preceding result and corollary (10.11), X is the topological direct sum of B and B^{d}.

(11.2) PROPOSITION. *Let $(X, C, \mathscr{P})$ be a topological Riesz space. Then the following statements hold:*

(a) *C is $\mathscr{P}$-closed;*

(b) *(X, C) is Archimedean;*

(c) *the solid hull of each $\mathscr{P}$-bounded set in X is $\mathscr{P}$-bounded, consequently if B is $\mathscr{P}$-bounded then so are B^+, B^-, and $|B|$, where*

$$B^+ = \{b^+ : b \in B\}, \qquad B^- = \{b^- : b \in B\}, \quad and \quad |B| = \{|b| : b \in B\};$$

(d) *C is a strict $\mathscr{B}$-cone in $(X, \mathscr{P})$;*

(e) *the $\mathscr{P}$-closure of each Riesz subspace of X is also a Riesz subspace.*

Proof. (a) Since $C = \{u \in X : u = u^+\}$, it follows from the continuity of the mapping $x \to x^+$ at 0 that C is $\mathscr{P}$-closed.

(b) follows from (a) and proposition (2.1).

(c) follows from the fact that $\mathscr{P}$ admits a neighbourhood-base at 0 consisting of solid sets in X.

(d) follows from (c).

(e) Obvious.

(11.3) PROPOSITION. *Let* $(X, C, \mathscr{P})$ *be a topological Riesz space. Then the following statements hold:*

(a) *the $\mathscr{P}$-closure $\bar{H}$ of each solid set H is solid;*

(b) *the solid kernel* sk(H) *of each $\mathscr{P}$-closed set H is $\mathscr{P}$-closed;*

(c) *H absorbs every $\mathscr{P}$-bounded set in X if and only if* sk(H) *absorbs all $\mathscr{P}$-bounded subsets of X, where $H \subset X$.*

Proof. (a) Let $\mathscr{U}$ be a neighbourhood-base at 0 for $\mathscr{P}$ consisting of solid sets in X. Since $\bar{H} = \cap\{H + V : V \in \mathscr{U}\}$, it follows from proposition (10.5)(e) that $\bar{H}$ is solid.

(b) Since sk(H) is the largest solid subset of X contained in H, it follows from (a) that sk(H) is $\mathscr{P}$-closed.

(c) The sufficiency is obvious, and the necessity follows from proposition (11.2)(c) and from the fact that sk(H) is the largest solid subset of X contained in H. This completes the proof.

(11.4) COROLLARY. *Let* $(X, C, \mathscr{P})$ *be a topological Riesz space. Then the $\mathscr{P}$-closure of each ℓ-ideal in X is an ℓ-ideal in X.*

In view of proposition (11.3), for any topological Riesz space $(X, C, \mathscr{P})$ there exists a neighbourhood-base at 0 for $\mathscr{P}$ consisting of $\mathscr{P}$-closed solid sets in X; in fact

$$\mathscr{U} = \{\bar{V} : V \text{ is a solid } \mathscr{P}\text{-neighbourhood of } 0\}$$

is such a neighbourhood-base at 0 for $\mathscr{P}$, where $\bar{V}$ is the $\mathscr{P}$-closure of V.

(11.5) PROPOSITION. *Let* $(X, C, \mathscr{P})$ *be a topological Riesz space. Then each normal subspace of X is $\mathscr{P}$-closed.*

Proof. Let B be any normal subspace of X. Since (X, C) is Archimedean, it follows from proposition (10.9) that $B = B^{\mathrm{dd}}$. The closedness of any set of the form A^{d} is an immediate consequence of the continuity of lattice operations, and hence B is $\mathscr{P}$-closed.

It should be noted that if B is an ℓ-ideal in a locally convex Riesz space $(X, C, \mathscr{P})$, then, by proposition (11.5), $\bar{B} \subset \{B\}$. Therefore the question arises whether $\bar{B} = \{B\}$; we shall see (in theorem (13.1)) that this is the case if and only if all continuous linear functionals are normal.

(11.6) PROPOSITION. *Let $(X, C, \mathscr{P})$ be a locally convex Riesz space. The topological completion $(\tilde{X}, \tilde{\mathscr{P}})$ of $(X, C, \mathscr{P})$ is a locally convex Riesz space ordered by the cone $\tilde{C}$, where $\tilde{C}$ is the $\tilde{\mathscr{P}}$-closure of C in $(\tilde{X}, \tilde{\mathscr{P}})$.*

Proof. Since the lattice operation $(x, y) \to x \vee y$ is uniformly continuous from $X \times X$ into $\tilde{X}$, it follows from a well-known result that this mapping has a unique uniformly continuous extension from $\tilde{X} \times \tilde{X}$ into $\tilde{X}$. It is easily verified that this extension makes $(\tilde{X}, \tilde{C})$ into a Riesz space. On the other hand, if the locally solid topology $\mathscr{P}$ is determined by a family $\{p_\tau : \tau \in D\}$ of Riesz semi-norms on (X, C), then, for each $\tau \in D$, p_τ can be uniquely extended to a continuous Riesz semi-norm $\tilde{p}_\tau$ on $(\tilde{X}, \tilde{C}, \tilde{\mathscr{P}})$, and $\tilde{\mathscr{P}}$ is determined by the family $\{\tilde{p}_\tau : \tau \in D\}$ of Riesz semi-norms on $(\tilde{X}, \tilde{C})$; it then follows from theorem (6.3) that $(\tilde{X}, \tilde{C}, \tilde{\mathscr{P}})$ is a locally convex Riesz space.

(11.7) PROPOSITION (Kawai). *Let $(X, C, \mathscr{P})$ be a locally convex Riesz space, and let (Y, K) be the order completion of (X, C). If $(X, C, \mathscr{P})$ satisfies the condition that $\{x_\gamma : \gamma \in I\}$ converges to 0 for $\mathscr{P}$ whenever $x_\gamma \downarrow 0$, then there exists a locally solid topology $\hat{\mathscr{P}}$ on (Y, K) such that $\hat{\mathscr{P}}$ induces $\mathscr{P}$ on (X, C) and X is dense in $(Y, K, \hat{\mathscr{P}})$.*

Proof. Let $\{p_\alpha : \alpha \in D\}$ be a family of Riesz semi-norms on X generating the locally solid topology $\mathscr{P}$. In view of the definition of order-completion, for each $y \in Y$ there is an increasing net $\{a_\lambda\}$ of X with $0 \leqslant \hat{a}_\lambda \leqslant |y|$ such that $|y| = \sup\{\hat{a}_\lambda\}$, we now define

$$\hat{p}_\alpha(x) = \sup_\lambda \{p_\alpha(a_\lambda)\} \qquad (\alpha \in D).$$

Since $(X, C, \mathscr{P})$ satisfies the property that $p_\alpha(x_\gamma) \to 0$ as $x_\gamma \downarrow 0$ $(\gamma \in D)$, then each $\hat{p}_\alpha$ is independent of the particular choice of an increasing net $\{a_\lambda\}$ with $\hat{a}_\lambda \uparrow |y|$. It is easily seen that each $\hat{p}_\alpha$ is a Riesz semi-norm on Y for which

$$\hat{p}_\alpha(\hat{a}) = p_\alpha(a) \qquad \text{for all } a \in X.$$

Therefore the family $\{\hat{p}_\alpha : \alpha \in D\}$ determines a locally solid topology $\hat{\mathscr{P}}$ on Y such that $\mathscr{P}$ is the relative topology on X induced by $\hat{\mathscr{P}}$, and that X is dense in $(Y, \hat{\mathscr{P}})$.

A Riesz semi-norm on X is called a *Riesz norm* if it is a norm. A Riesz space equipped with a Riesz norm is called a *normed Riesz space;* and a Riesz space with a Riesz semi-norm is called a *semi-normed Riesz space.* Obviously, normed Riesz spaces are locally convex Riesz spaces. A normed Riesz space is called a *Banach lattice* (or *B-lattice*) if it is complete for the norm. A locally convex Riesz space $(X, C, \mathscr{P})$ is called a *Fréchet lattice* (or *F-lattice*) if it is metrizable and complete for the topology $\mathscr{P}$. Similar to the case of locally convex spaces, we shall see (below) that each locally convex Riesz space $(X, C, \mathscr{P})$ is ℓ-isomorphic and topologically isomorphic with a Riesz subspace Y of the projective limit of B-lattices $\{L_\alpha : \alpha \in \Gamma\}$.

It is easily seen that every Riesz subspace of a locally convex Riesz space, equipped with the relative topology, is a locally convex Riesz space; and that the Cartesian product of a family of locally convex Riesz spaces, equipped with the product topology, is a locally convex Riesz space. If J is a closed ℓ-ideal in a locally convex Riesz space $(X, C, \mathscr{P})$ then the quotient space $(X/J, [C]_J)$, equipped with the quotient topology, is also a locally convex Riesz space; for the sake of convenience, it is referred to as the *quotient Riesz space.* We shall see that the locally convex direct sum of a family of locally convex Riesz spaces is also a locally convex Riesz space.

(11.8) PROPOSITION. *Every locally convex Riesz space $(X, C, \mathscr{P})$ is ℓ-isomorphic and topologically isomorphic to a dense Riesz subspace of the projective limit of a family of Banach lattices; this family can be so chosen that its cardinality equals the cardinality of a given neighbourhood-base at 0 for $\mathscr{P}$.*

Proof. Let $\{p_\alpha : \alpha \in \Gamma\}$ be a family of continuous Riesz semi-norms on $(X, C, \mathscr{P})$ generating $\mathscr{P}$. Γ is a directed set when we define $\alpha \leqslant \beta$ if $p_\alpha(x) \leqslant p_\beta(x)$ for all $x \in X$. For each $\alpha \in \Gamma$, let

$$J_\alpha = \{x \in X : p_\alpha(x) = 0\};$$

then J_α is a $\mathscr{P}$-closed ℓ-ideal in X. Further, we define

$$[C]_\alpha = \{[x]_\alpha \in X/J_\alpha : \exists j_\alpha \in J_\alpha \text{ such that } x + j_\alpha \in C\},$$

$$\|[x]_\alpha\|_\alpha = p_\alpha(x),$$

then $(X/J_\alpha, [C]_\alpha, \|\cdot\|_\alpha)$ is a normed Riesz space and the norm topology $\|\cdot\|_\alpha$ is coarser than the quotient topology on X/J_α induced by $\mathscr{P}$. Moreover, a continuous ℓ-homomorphism $\hat{g}_{\alpha\beta}$ from $(X/J_\beta, [C]_\beta, \|\cdot\|_\beta)$ onto $(X/J_\alpha, [C]_\alpha, \|\cdot\|_\alpha)$ is defined by setting $[x]_\alpha = \hat{g}_{\alpha\beta}([x]_\beta)$ whenever $\alpha \leqslant \beta$. By proposition (11.6), the completion L_α of $(X/J_\alpha, [C]_\alpha, \|\cdot\|_\alpha)$ is a Banach lattice ($\alpha \in \Gamma$), and hence $\hat{g}_{\alpha\beta}$ can be uniquely extended to a continuous ℓ-homomorphism $g_{\alpha\beta}$ from L_β into L_α whenever $\alpha \leqslant \beta$. If π_α denotes the projection of $\prod_{\alpha \in \Gamma} L_\alpha$ into L_α, then the projective limit $\varprojlim g_{\alpha\beta}L_\beta$ of the family of Banach lattices $\{L_\alpha : \alpha \in \Gamma\}$ with respect to the ℓ-homomorphisms $g_{\alpha\beta}(\alpha, \beta \in \Gamma, \alpha \leqslant \beta)$ is a closed ℓ-ideal in the product space $\prod_\alpha L_\alpha$, because

$$\varprojlim g_{\alpha\beta}L_\beta = \{z \in \prod_\alpha L_\alpha : (\pi_\alpha - g_{\alpha\beta} \circ \pi_\beta)(z) = 0, \quad \alpha \leqslant \beta, \alpha, \beta \in \Gamma\}$$

and $\pi_\alpha - g_{\alpha\beta} \circ \pi_\beta$ is a continuous ℓ-homomorphism from $\prod_\alpha L_\alpha$ into L_α. Notice also that the product space $\prod_\alpha L_\alpha$ is complete with respect to the product topology $\mathscr{I}$; it then follows that $\varprojlim g_{\alpha\beta}L_\beta$ is a complete locally convex Riesz space with respect to the relative topology $\mathscr{I}_r$ induced by $\mathscr{I}$.

Now if we define a mapping ψ from X into $\prod_\alpha L_\alpha$ by setting

$$\psi(x) = ([x]_\alpha : \alpha \in \Gamma) \quad (x \in X),$$

then ψ is an ℓ-isomorphism and a topological isomorphism from $(X, C, \mathscr{P})$ into $(\varprojlim g_{\alpha\beta}L_\beta, \mathscr{I}_r)$, and $\psi(X)$ and $\prod_\alpha X/J_\alpha$ are ℓ-isomorphic; this implies that $\psi(X)$ is a dense Riesz subspace of $\left(\varprojlim g_{\alpha\beta}L_\beta, \mathscr{I}_r\right)$.

We now turn our attention to seek some conditions ensuring that the inductive topology with respect to a family of locally convex Riesz spaces is locally solid.

(11.9) PROPOSITION. *Let (X, C) be a Riesz space, and let $\mathscr{P}$ be the inductive topology on X with respect to locally convex Riesz spaces $\{(X_\alpha, C_\alpha, \mathscr{P}_\alpha) : \alpha \in \Gamma\}$ and ℓ-homomorphisms $\{T_\alpha : \alpha \in \Gamma\}$. If X is the linear hull of $\bigcup_{\alpha \in \Gamma} T_\alpha(X_\alpha)$ and if each $T_\alpha(X_\alpha)$ is an ℓ-ideal in X, then $\mathscr{P}$ is a locally solid topology and hence $(X, C, \mathscr{P})$ is a locally convex Riesz space.*

Proof. Let V be any convex $\mathscr{P}$-neighbourhood of 0 in X. Then each $T_\alpha^{-1}(V)$ is a $\mathscr{P}_\alpha$-neighbourhood of 0 in X_α, and so there exists a solid and convex $\mathscr{P}_\alpha$-neighbourhood W_α of 0 in X_α such that $W_\alpha \subset T_\alpha^{-1}(V)$. By proposition (10.24)(e), each $T_\alpha(W_\alpha)$ is a solid subset of $T_\alpha(X_\alpha)$, and hence $T_\alpha(W_\alpha)$ is a solid subset of X because each $T_\alpha(X_\alpha)$ is an ℓ-ideal in X. Observe that the union of a family of solid sets in X is a solid subset of X. It follows that $W = \bigcup_{\alpha \in \Gamma} T_\alpha(W_\alpha)$ is a solid subset of X, and hence, from proposition (10.5)(a), the convex hull U of W is a solid subset of X. Since X is the linear hull of $\bigcup_{\alpha \in \Gamma} T_\alpha(X_\alpha)$, U is absorbing, and so U is a convex, solid, and absorbing subset of X. It is clear that U is a $\mathscr{P}$-neighbourhood of 0 since $W_\alpha \subset T_\alpha^{-1}(W) \subset T_\alpha^{-1}(U)$, and that $U \subseteq V$ since V is a convex set containing W. Therefore we have found a convex, solid $\mathscr{P}$-neighbourhood U of 0 in X such that $U \subset V$; consequently $\mathscr{P}$ is a locally solid topology, and this completes the proof.

In view of the definition of a strict inductive limit and the preceding result, it is known that if $\{X_n\}$ is a sequence of ℓ-ideals in X, and if $(X_n, C_n, \mathscr{P}_n)$ are locally convex Riesz spaces, then the strict inductive limit of $\{X_n\}$ is a locally convex Riesz space.

(11.10) COROLLARY. *The locally convex direct sum of a family of locally convex Riesz spaces* $\{(X_\alpha, C_\alpha, \mathscr{P}_\alpha): \alpha \in \Gamma\}$, *denoted by* $\bigoplus_\alpha (X_\alpha, C_\alpha, \mathscr{P}_\alpha)$, *is a locally convex Riesz space. Furthermore,* $\bigoplus_\alpha (X_\alpha, C_\alpha, \mathscr{P}_\alpha)$ *is complete if and only if each* X_α *is complete for* $\mathscr{P}_\alpha$.

Proof. If X denotes the algebraic direct sum of $\{X_\alpha: \alpha \in \Gamma\}$, j_α denotes the injection map of X_α into X, and if $C = X \cap \prod_{\alpha \in \Gamma} C_\alpha$, then the Riesz space (X, C) is the linear hull of $\bigcup_{\alpha \in \Gamma} j_\alpha(X_\alpha)$, and each $j_\alpha(X_\alpha)$ is an ℓ-ideal in X. The result now follows from proposition (11.9) and the definition of locally convex sum topology on X.

Let (E, C) be an ordered vector space, and let e be an order-unit in E. Then the gauge $\|\cdot\|_e$ of $[-e, e]$, defined by

$$\|x\|_e = \inf\{\lambda > 0 : x \in \lambda[-e, e]\},$$

is a semi-norm on E, and it is referred to as the *order-unit semi-norm* corresponding to e. It is clear that $\|\cdot\|_e$ is a norm if and only if (E, C) is almost-Archimedean, and that the order-unit semi-norms corresponding to two different order-units are equivalent; so the topology given by order-unit semi-norms can be regarded as a topology determined by the ordering. An order-unit semi-norm, which is also a norm, is called an *order-unit norm*.

A normed Riesz space $(X, C, \|\,.\,\|)$ is called a *unital normed Riesz space* if the norm $\|\,.\,\|$ is an order-unit norm. It is clear that if $(X, C, \|\,.\,\|)$ is a unital normed Riesz space, then the norm topology is the order-bound topology. This leads to the following result.

(11.11) PROPOSITION. *Let $(X, C, \mathscr{P})$ be a locally convex Riesz space. Then $\mathscr{P}$ is the order-bound topology $\mathscr{P}_{\mathrm{b}}$ if and only if there exists a family of unital normed Riesz spaces $\{(X_u, C_u, \|\,.\,\|_u) : u \in \Gamma\}$ and a family of ℓ-homomorphisms $\{T_u : u \in \Gamma\}$ with the following properties:*
  (a) *X is the linear hull of $\bigcup_{u \in \Gamma} T_u(X_u)$;*
  (b) *each $T_u(X_u)$ is an ℓ-ideal in X;*
  (c) *$\mathscr{P}$ is the inductive topology of $\{(X_u, C_u, \|\,.\,\|_u) : u \in \Gamma\}$ with respect to $\{T_u : u \in \Gamma\}$.*

Proof. (i) *Necessity.* For each $u \in C$, let

$$X_u = \bigcup_n n[-u, u],$$

$C_u = X_u \cap C$, $\|\,.\,\|_u$ be the gauge of $[-u, u]$ on X_u, and let T_u be the injection of X_u into X. Then each $(X_u, C_u, \|\,.\,\|_u)$ is a unital normed Riesz space with the order-unit u, $X = \bigcup_{u \in C} X_u$, and each X_u is an ℓ-ideal in X. Suppose that $\mathscr{T}$ is the inductive topology of

$$\{(X_u, C_u, \|\,.\,\|_u) : u \in C\}$$

with respect to $\{T_u, u \in C\}$. Then, by proposition (11.9), $\mathscr{T}$ is locally solid, and *a fortiori* $\mathscr{T}$ is coarser than $\mathscr{P}_{\mathrm{b}}$. On the other hand, since the relative topology on each X_u induced by $\mathscr{P}$ is coarser than the norm topology $\|\,.\,\|_u$, it follows that the injection map T_u from $(X_u, C_u, \|\,.\,\|_u)$ into $(X, C, \mathscr{P})$ is continuous, and hence from the definition of inductive topology that $\mathscr{P}$ is coarser than $\mathscr{T}$. Therefore, $\mathscr{P}_{\mathrm{b}}$ is the inductive topology of $\{(X_u, C_u, \|\,.\,\|_u) : u \in C\}$ with respect to $\{T_u : u \in C\}$.

(ii) *Sufficiency.* Let $\|\,.\,\|_u$ be the order-unit norm on X_u corresponding to the order-unit $u \in X_u$. In order to show that $\mathscr{P}$ is the order-bound topology $\mathscr{P}_{\mathrm{b}}$, it is sufficient to verify that each convex and circled subset V of X, which absorbs all order-bounded subsets of X, is a $\mathscr{P}$-neighbourhood of 0 or, equivalently, each $T_u^{-1}(V)$ is a $\|\,.\,\|_u$-neighbourhood of 0 in X_u. Let V be such a set. Since u is an order-unit in X_u and since $\|.\|_u$ is the order-unit norm corresponding to u, it follows that the order-interval $[-u, u]$ is a $\|.\|_u$-neighbourhood of 0 in X_u. Notice that $T_u([-u, u]) = [-T_u(u), T_u(u)]$ is an order-interval in X; then, there exists a $\lambda > 0$ such that $[-T_u(u), T_u(u)] \subset \lambda V$. We conclude from $[-u, u] \subset \lambda T_u^{-1}(V)$ that $T_u^{-1}(V)$ is a $\|.\|_u$-neighbourhood of 0 in X_u; hence $\mathscr{P}$ is the order-bound topology $\mathscr{P}_b$.

We shall now turn our attention to the topological completeness. It is natural to ask under what conditions an order-complete topological Riesz space is necessarily topologically complete.

(11.12) DEFINITION. A topological Riesz space $(X, C, \mathscr{P})$ is said to be *locally order-complete* if there exists a neighbourhood-base at 0 for $\mathscr{P}$ consisting of solid and order-complete sets in X.

Obviously each locally order-complete topological Riesz space must be order-complete.

(11.13) PROPOSITION (Nakano). *Let $(X, C, \mathscr{P})$ be a topological Riesz space. If $(X, C, \mathscr{P})$ is locally order-complete, then each order-interval in X is complete for $\mathscr{P}$.*

Proof (Schaefer). Since X is a Riesz space, it is enough to verify that each order-interval of the form $[0, u]$ $(u \in C)$ is $\mathscr{P}$-complete. We first prove that the result holds for the special case when $\mathscr{P}$ is metrizable. Choose a countable neighbourhood-base $\{W_n : n \in \mathbf{N}\}$ at 0 for $\mathscr{P}$ consisting of solid order-complete sets and $W_{n+1} + W_{n+1} \subset W_n$ for all n. Any $\mathscr{P}$-Cauchy sequence in $[0, u]$ has a subsequence $\{x_n\}$ such that

$$x_{n+1} - x_n \in W_{n+1} \quad \text{for all} \quad n \in \mathbf{N}.$$

Suppose that

$$y_r = \sup\{x_n : n \geqslant r\}, \qquad z_r = \inf\{x_n : n \geqslant r\}.$$

For fixed r and for any natural number q with $q \geqslant r$,

$$\sup\{x_n : r \leqslant n \leqslant q\} - x_r$$

is increasing with respect to q and

$$0 \leqslant \sup\{x_n - x_r : r \leqslant n \leqslant q\} \leqslant u;$$

it follows from the order-completeness of X that

$$w = \sup_{q \geqslant r}\{\sup\{x_n - x_r : r \leqslant n \leqslant q\}\}$$

exists in X. Furthermore we also have

$$w = \sup_{q \geqslant r}\{\sup\{x_n : r < n < q\}\} - x_r$$
$$= \sup\{x_n : n > r\} - x_r = y_r - x_r.$$

On the other hand, notice that

$$0 \leqslant \sup\{x_n : r < n < q\} - x_r \leqslant \sup\{|x_n - x_r| : r < n < q\}$$
$$\leqslant |x_{r+1} - x_r| + |x_{r+2} - x_{r+1}| + \ldots + |x_q - x_{q-1}|$$
$$\in W_{r+1} + W_{r+2} + \ldots + W_q \subset W_r.$$

It then follows from the order-completeness of W_r that $w \in W_r$, and hence that $y_r \in x_r + W_r$. A similar argument shows that $z_r \in x_r + W_r$. Therefore we have, for any $r \in \mathbf{N}$,

$$y_r \in x_r + W_r \quad \text{and} \quad z_r \in x_r + W_r. \tag{11.1}$$

It is clear that $y_r \downarrow (z_r \uparrow)$ and that $0 \leqslant y_r (z_r \leqslant u)$ for all r; then

$$y = \inf\{y_r : r > 1\}, \qquad z = \sup\{z_r : r > 1\}$$

exist in X, and y, z are in $[0, u]$. Since

$$y - z = (y - x_{n+1}) - (z - x_{n+1}) \leqslant (y_{n+1} - x_{n+1}) - (z_{n+1} - x_{n+1})$$
$$\in W_{n+1} + W_{n+1} \subset W_n$$

for all n, we conclude from $\bigcap_{n \in \mathbf{N}} W_n = \{0\}$ that $y - z = 0$; hence

$$\inf\{y_r : r > 1\} = \sup\{z_r : r > 1\} = y. \tag{11.2}$$

We now claim that x_n converges to y with respect to $\mathscr{P}$. Indeed we note that

$$z_n - x_n \leqslant z - x_n = y - x_n \leqslant y_n - x_n.$$

Since $z_n - x_n$ and $y_n - x_n$ converge to 0 and $\mathscr{P}$ is locally solid (so locally order-convex) it follows from theorem (5.1) that x_n converges to y with respect to $\mathscr{P}$. This proves the result for the metrizable case.

Now let us consider the general case, that is the case when $\mathscr{P}$ is not necessarily metrizable. Let $\mathscr{U} = \{W_\alpha : \alpha \in D\}$ be a neighbourhood-base at 0 for $\mathscr{P}$ consisting of solid and order-complete sets in X. Define Ω to be the class of all countable collections

$$\mathscr{C} = \{W_{\alpha_n} : \alpha_n \in D, n = 1, 2 \ldots\}$$

of $\mathscr{U}$ each of which $\mathscr{C}$ forms a neighbourhood-base at 0 for a vector topology, $\mathscr{P}_\mathscr{C}$ say. Then Ω is non-empty, directed by inclusion,

$$\mathscr{U} = \cup \{\mathscr{C} : \mathscr{C} \in \Omega\},$$

and $(X, C, \mathscr{P}_{\mathscr{C}})$ is a pseudo-metrizable (not necessarily Hausdorff) locally order-complete topological Riesz space. For each $\alpha \in D$, let

$$W_\alpha^{(0)} = \bigcap_{\lambda > 0} \lambda W_\alpha,$$

and suppose that

$$N_0 = \cap \{W_\alpha^{(0)} : \alpha \in D\}, \qquad N_{\mathscr{C}} = \cap \{W_{\alpha_n}^{(0)} : W_{\alpha_n} \in \mathscr{C}\}.$$

Then N_0 and $N_{\mathscr{C}}$ are normal subspaces of X. Since X is order-complete, then, by corollary (10.11), we have, for each $\mathscr{C} \in \Omega$,

$$X = N_{\mathscr{C}} \oplus N_{\mathscr{C}}^d = N_0 \oplus N_0^d.$$

Let P be the ℓ-projection of X onto $N_{\mathscr{C}}^d$, and let P_0 be the ℓ-projection of X onto N_0^d. The following conclusions concerning ℓ-projections are easily verified:

(a) if $\mathscr{C}_1 \subset \mathscr{C}_2$ then $P_{\mathscr{C}_1} \leqslant P_{\mathscr{C}_2}$ and $P_{\mathscr{C}_1} \circ P_{\mathscr{C}_2} = P_{\mathscr{C}_1}$;

(b) $0 \leqslant P_{\mathscr{C}} \leqslant P_0 = I$, where I is the identity map;

(c) $P_0 = \sup\{P_{\mathscr{C}} : \mathscr{C} \in \Omega\}$;

(d) if $u_\tau \uparrow u$ then $P_{\mathscr{C}}(u_\tau) \uparrow P_{\mathscr{C}}(u)$ for each $\mathscr{C} \in \Omega$;

(e) if $\mathscr{P}_{\mathscr{C}}^d$ denotes the vector topology on $N_{\mathscr{C}}^d$ induced by $\mathscr{P}_{\mathscr{C}}$, then $P_{\mathscr{C}}$ is a continuous ℓ-projection of $(X, C, \mathscr{P})$ into $(N_{\mathscr{C}}^d, \mathscr{P}_{\mathscr{C}}^d)$.

Suppose that $\{x_\alpha\}$ is a $\mathscr{P}$-Cauchy net in $[0, u]$. For each $\mathscr{C} \in \Omega$, $\{\mathscr{P}_{\mathscr{C}}(x_\alpha)\}$ is a $\mathscr{P}_{\mathscr{C}}^d$-Cauchy net in $N_{\mathscr{C}}^d$; since $(N_{\mathscr{C}}^d, \mathscr{P}_{\mathscr{C}}^d)$ is a metrizable locally order-complete topological Riesz space, there exists

$$y_{\mathscr{C}} \in [0, P_{\mathscr{C}}(u)]$$

such that $P_{\mathscr{C}}(x_\alpha)$ converges to $y_{\mathscr{C}}$ with respect to $\mathscr{P}_{\mathscr{C}}^d$; for convenience, it is denoted by $y_{\mathscr{C}} = \mathscr{P}_{\mathscr{C}}^d - \lim_\alpha P_{\mathscr{C}}(x_\alpha)$. By (b), $y_{\mathscr{C}} \in [0, u]$. Therefore we find a net $\{y_{\mathscr{C}} : \mathscr{C} \in \Omega\}$ in $[0, u]$, where

$$y_{\mathscr{C}} = \mathscr{P}_{\mathscr{C}}^d - \lim_\alpha P_{\mathscr{C}}(x_\alpha).$$

Moreover, the net $\{y_{\mathscr{C}} : \mathscr{C} \in \Omega\}$ has the following properties:

(i) if $\mathscr{C}_1 \subset \mathscr{C}_2$ then $y_{\mathscr{C}_1} = \mathscr{P}_{\mathscr{C}_2}^d - \lim_\alpha P_{\mathscr{C}_1}(x_\alpha)$,

(ii) $v = \sup\{y_{\mathscr{C}} : \mathscr{C} \in \Omega\}$ exists in X and

$$P_{\mathscr{C}_1}(v) = y_{\mathscr{C}_1}, \quad \text{for any } \mathscr{C}_1 \in \Omega.$$

In fact, if $\mathscr{C}_1 \subset \mathscr{C}_2$, then $N_{\mathscr{C}_1}^d \subset N_{\mathscr{C}_2}^d$, and so property (i) follows from the fact that $(N_{\mathscr{C}}^d, \mathscr{P}_{\mathscr{C}}^d)$ is Hausdorff. To prove property (ii), we observe that $0 \leqslant y_{\mathscr{C}} \uparrow \leqslant u$; it follows from the order-completeness of X that

$$v = \sup\{y_{\mathscr{C}} : \mathscr{C} \in \Omega\}$$

exists in X $(0 \leqslant v \leqslant u)$. On the other hand, if $\mathscr{C}_1 \subset \mathscr{C}_2$ then, by (a), (e), and property (i), we have

$$P_{\mathscr{C}_1}(y_{\mathscr{C}_2}) = P_{\mathscr{C}_1}(\mathscr{P}^d_{\mathscr{C}_2} - \lim_\alpha P_{\mathscr{C}_2}(x_\alpha)) = P^d_{\mathscr{C}_2} - \lim_\alpha (P_{\mathscr{C}_1} \circ P_{\mathscr{C}_2})(x_\alpha)$$

$$= P^d_{\mathscr{C}_2} - \lim_\alpha P_{\mathscr{C}_1}(x_\alpha) = y_{\mathscr{C}_1}.$$

We conclude from (d) that

$$P_{\mathscr{C}_1}(v) = \sup\{P_{\mathscr{C}_1}(y_{\mathscr{C}}) : \mathscr{C} \in \Omega\} = \sup\{P_{\mathscr{C}_1}(y_{\mathscr{C}}) : \mathscr{C}_1 \subset \mathscr{C} \in \Omega\} = y_{\mathscr{C}_1},$$

verifying property (ii).

Therefore we complete the proof by showing that x_α converges to v with respect to $\mathscr{P}$. In fact, since

$$P_{\mathscr{C}}(v) = y_{\mathscr{C}} = P^d_{\mathscr{C}} - \lim_\alpha P_{\mathscr{C}}(x_\alpha) \qquad (\mathscr{C} \in \Omega),$$

$P_{\mathscr{C}}(v)$ is a $\mathscr{P}^d_{\mathscr{C}}$-cluster point of $\{P_{\mathscr{C}}(x_\alpha)\}$. Note also that $P_{\mathscr{C}}$ is a continuous ℓ-projection of $(X, \mathscr{P}_{\mathscr{C}})$ onto $(N^d_{\mathscr{C}}, \mathscr{P}^d_{\mathscr{C}})$ which vanishes on $N_{\mathscr{C}}$; whence v is a $\mathscr{P}_{\mathscr{C}}$-cluster point of $\{x_\alpha\}$. Clearly $\{x_\alpha\}$ is a $\mathscr{P}_{\mathscr{C}}$-Cauchy net for each $\mathscr{C} \in \Omega$; consequently x_α converges to v with respect to $\mathscr{P}_{\mathscr{C}}$ for each $\mathscr{C} \in \Omega$. We conclude from

$$\mathscr{U} = \cup \{\mathscr{C} : \mathscr{C} \in \Omega\}$$

that x_α converges to v with respect to $\mathscr{P}$, and hence that $[0, u]$ is $\mathscr{P}$-complete.

Recall that an ordered topological vector space $(E, C, \mathscr{P})$ is said to be *boundedly order-complete* if each increasing net in E which is $\mathscr{P}$-bounded has a supremum in E. We now present one of the deepest results in the theory of topological Riesz spaces concerning the topological completeness.

(11.14) THEOREM (Nakano). *A topological Riesz space $(X, C, \mathscr{P})$ which is both locally order-complete and boundedly order-complete is complete for $\mathscr{P}$.*

Proof. Let $\{x_\tau : \tau \in D\}$ be a $\mathscr{P}$-Cauchy net. In view of proposition (11.1), $\{x_\tau^+ : \tau \in D\}$ and $\{x_\tau^- : \tau \in D\}$ are also $\mathscr{P}$-Cauchy nets, and hence it is sufficient to verify that $\{x_\tau^+ : \tau \in D\}$ is convergent with respect to $\mathscr{P}$ since the convergence of $\{x_\tau\}$ follows immediately from that of $\{x_\tau^+\}$ and $\{x_\tau^-\}$.

For any $x \in C$, the continuity of lattice operations ensures that $\{x_\tau^+ \wedge x : \tau \in D\}$ is a $\mathscr{P}$-Cauchy net in the order-interval $[0, x]$; it then follows from proposition (11.13) that there exists $A_x \in [0, x]$ such that $x_\tau^+ \wedge x$ converges to A_x with respect to $\mathscr{P}$. Therefore we have found a net $\{A_x : x \in C\}$ in C, where A_x is the limit of $x_\tau^+ \wedge x$ with respect to $\mathscr{P}$. We now claim that $\{A_x : x \in C\}$ is directed upwards and $\mathscr{P}$-bounded. For any $x, y \in C$, $x \vee y \in C$, by making use of the continuity of lattice operations, we have

$$A_{x \vee y} = \lim_\tau \{x_\tau^+ \wedge (x \vee y)\} = \{\lim_\tau x_\tau^+ \wedge x\} \vee \{\lim_\tau x_\tau^+ \wedge y\} = A_x \vee A_y$$

so that $\{A_x : x \in C\}$ is directed upwards. On the other hand, let W be any solid $\mathscr{P}$-neighbourhood of 0, and let V be a solid $\mathscr{P}$-neighbourhood of 0 such that $V + V + V \subset W$. Since $\{x_\tau^+ : \tau \in D\}$ is a $\mathscr{P}$-Cauchy net, there exists $\tau_0 \in D$ such that

$$x_\tau^+ - x_{\tau'}^+ \in V \quad \text{whenever} \quad \tau, \tau' \geqslant \tau_0;$$

for this τ_0 there exists λ_0 with $\lambda_0 \geqslant 1$ such that $x_{\tau_0}^+ \in \lambda_0 V$, and so

$$x_\tau^+ = (x_\tau^+ - x_{\tau_0}^+) - x_{\tau_0}^+ \in V + \lambda_0 V \subset \lambda_0 (V + V).$$

Since $x_\tau^+ \wedge x$ converges to A_x with respect to $\mathscr{P}$, there exists $\tilde{\tau}_0 \in D$ such that

$$A_x - x_\tau^+ \wedge x \in V \quad \text{whenever} \quad \tau \geqslant \tilde{\tau}_0.$$

Take $\tau_1 \in D$ such that $\tau_1 \geqslant \tau_0$ and $\tau_1 \geqslant \tilde{\tau}_0$. We conclude, from

$$0 \leqslant A_x = (A_x - x_\tau^+ \wedge x) + x_\tau^+ \wedge x \leqslant (A_x - x_\tau^+ \wedge x) + x_\tau^+ \in V + \lambda_0 (V + V)$$

$$\subset \lambda_0 (V + V + V) \subset \lambda_0 W \quad \text{whenever} \quad \tau \geqslant \tau_1,$$

that $\{A_x : x \in C\}$ is $\mathscr{P}$-bounded.

According to the hypothesis,

$$a = \sup\{A_x : x \in C\}$$

exists in X. Furthermore, the element a has the following property:

$$a \wedge y = A_y \quad \text{for any} \quad y \in C. \tag{11.2}$$

In fact, by proposition (10.2), we obtain

$$a \wedge y = y \wedge \sup\{A_x : x \in C\} = \sup\{y \wedge A_x : x \in C\};$$

since

$$y \wedge A_x = y \wedge \{\lim_\tau x_\tau^+ \wedge x\} = \lim_\tau (x_\tau^+ \wedge x \wedge y) = A_{x \wedge y} \quad (x \in C)$$

and since $A_{x \wedge y} \leqslant A_y$ for all $x \in C$, it follows that

$$a \wedge y = A_y.$$

We complete the proof by showing that x_τ^+ converges to a with respect to $\mathscr{P}$. Let W be any $\mathscr{P}$-neighbourhood of 0, and let V be a solid $\mathscr{P}$-neighbourhood of 0 such that $V+V \subset W$. Since $\{x_\tau^+ : \tau \in D\}$ is a $\mathscr{P}$-Cauchy net, there exists $\tau_0 \in D$ such that

$$x_\tau^+ - x_{\tau'}^+ \in V \quad \text{whenever} \quad \tau, \tau' \geqslant \tau_0. \tag{11.3}$$

On the other hand, for any $\tau_1 \in D$ with $\tau_1 \geqslant \tau_0$, let

$$x_1 = x_{\tau_1}^+ \vee a$$

then, by equality (11.2),

$$A_{x_1} = a \wedge x_1 = a \wedge (x_{\tau_1}^+ \vee a) = a.$$

Notice also that $x_\tau^+ \wedge x_1$ converges to A_{x_1} with respect to $\mathscr{P}$; then there exists $\tau_2 \in D$ such that

$$x_\tau^+ \wedge x_1 - a = x_\tau^+ \wedge x_1 - A_{x_1} \in V \quad \text{whenever} \quad \tau \geqslant \tau_2. \tag{11.4}$$

Since

$$x_{\tau_1}^+ - a = (x_{\tau_1}^+ - x_\tau^+ \wedge x_1) + (x_\tau^+ \wedge x_1 - a) \quad (\tau \in D) \tag{11.5}$$

and since

$$|x_{\tau_1}^+ - x_\tau^+ \wedge x_1| = |x_{\tau_1}^+ \wedge x_1 - x_\tau^+ \wedge x_1| \leqslant |x_{\tau_1}^+ - x_\tau^+| \quad (\tau \in D), \tag{11.6}$$

it then follows from formulae (11.3), (11.4), (11.5), and (11.6) that

$$x_{\tau_1}^+ - a \in V + V \subset W,$$

where $\tau_1 \geqslant \tau_0$. This implies that

$$x_\tau^+ - a \in W \quad \text{whenever} \quad \tau \geqslant \tau_0,$$

and hence $\{x_\tau^+ : \tau \in D\}$ converges to a with respect to $\mathscr{P}$.

It should be noted that the above result is still true whenever $\mathscr{P}$ is non-Hausdorff. For a proof, see, for instance, Peressini (1967).

The following example shows that a topological Riesz space which is topologically complete need not be boundedly order-complete.

(11.15) EXAMPLE. Consider the space c_0 consisting of all null sequences of real numbers with the usual ordering and norm. Then c_0 is a B-lattice. Let $e_n = (1, 1, \ldots 1, 0, 0 \ldots) \in c_0$ and let $B = \{e_n : n \in \mathbf{N}\}$. Then B is increasing and norm-bounded. It is obvious that B does not have a supremum (not even any upper bound) in c_0. This shows that c_0 is not boundedly order-complete.

It is known from corollary (6.5) that the topological dual of any locally convex Riesz space $(X, C, \mathscr{P})$ is an ℓ-ideal in X^b. Therefore it is natural to ask whether the strong dual of a locally convex Riesz

11

space is also a locally convex Riesz space. We give an affirmative answer of this question as follows.

(11.16) THEOREM. *Let $(X, C, \mathscr{P})$ be a locally convex Riesz space with the topological dual X', and let $\beta(X', X)$ be the strong topology on X'. Then $(X', C', \beta(X', X))$ is a locally convex Riesz space and locally order-complete. If, in addition, $(X, \mathscr{P})$ is infrabarrelled, then $(X', C', \beta(X', X))$ is boundedly order-complete and hence X' is complete for $\beta(X', X)$.*

Proof. Since the polar, taken in X', of any solid set in X is solid in X', and since the solid hull of each $\mathscr{P}$-bounded set in X is $\mathscr{P}$-bounded, then $(X', C', \beta(X', X))$ is an order-complete locally convex Riesz space. Suppose that $\mathscr{B}$ is the family consisting of all convex, solid, $\mathscr{P}$-bounded subsets of X. Then $\mathscr{U} = \{B^0 : B \in \mathscr{B}\}$ is a neighbourhood-base at 0 for $\beta(X', X)$, where B^0 is the polar of B taken in X'. We now show that each B^0 is order-complete. Let $\{f_\tau\}$ be an increasing net in B^0 which is majorized in X'. Then $f = \sup f_\tau$ exists in X' because (X', C') is clearly order-complete. For any $x \in B$, we obtain

$$f(x) \leqslant f(|x|) = \sup f_\tau(|x|) \leqslant 1.$$

Therefore $f \in B^0$, and hence each B^0 is order-complete; consequently $(X', C', \beta(X', X))$ is locally order-complete.

Suppose now that $(X, \mathscr{P})$ is infrabarrelled, and that $\{f_\tau\}$ is a $\beta(X', X)$-bounded subset of X' which is directed upwards. Then, by the Alaoglu–Bourbaki theorem, $\{f_\tau\}$ is relatively $\sigma(X', X)$-compact, and hence $\{f_\tau\}$ has a $\sigma(X', X)$-cluster point f in X'. Since $f_\tau \uparrow$ and since C' is $\sigma(X', X)$-closed, it follows that $f = \sup f_\tau$ in X'. Therefore $(X', C', \beta(X', X))$ is boundedly order-complete. Finally, in view of the Nakano theorem (11.14), X' is complete for $\beta(X', X)$. Therefore the proof of the theorem is complete.

We shall see in the next chapter that $(X', C', \sigma_S(X', X))$ is also locally order-complete for any locally convex Riesz space $(X, C, \mathscr{P})$, where $\sigma_S(X', X)$ is the locally solid topology associated with $\sigma(X', X)$.

(11.17) DEFINITION. *Let $(X, C, \mathscr{P})$ be a locally convex Riesz space with the topological dual X'. Then $(X', C', \beta(X', X))$ is called the Riesz dual of X and $(X'', C'', \beta(X'', X'))$ is referred to as the Riesz bidual of X, where $X'' = (X', \beta(X', X))'$ and C'' is the dual cone of C'.*

(11.18) PROPOSITION. *Let $(X, C, \mathscr{P})$ be a locally convex Riesz space with the Riesz dual $(X', C', \beta(X', X))$. Then the image $\hat{X}$ of X into X'' under the evaluation map $x \to \hat{x}$, defined by*

$$\hat{x}(f) = f(x) \quad \text{for all} \quad f \in X',$$

is a Riesz subspace of X'', and the evaluation map is an ℓ-isomorphism of X onto $\hat{X}$. Furthermore, $\hat{X} \subset (X')_n^b$, and hence $\hat{X} \subset X'' \cap (X')_n^b$.

Proof. It is clear that the evaluation map is a bijection of X onto $\hat{X}$. We now show that $\widehat{x^+} = (\hat{x})^+$. Let f be in C'. It follows from proposition (10.15)(a) that

$$\widehat{x^+}(f) = f(x^+) = \sup\{g(x): 0 \leqslant g \leqslant f\}.$$

Obviously $\hat{x}$ is an order-bounded linear functional on X'; in view of theorem (10.12)(a), we obtain

$$(\hat{x})^+(f) = \sup\{\hat{x}(g): 0 \leqslant g \leqslant f\}.$$

Therefore $\widehat{x^+}(f) = (\hat{x})^+(f)$, hence $\widehat{x^+} = (\hat{x})^+$ because of $X' = C' - C'$, and thus the evaluation map $x \to \hat{x}$ is an ℓ-isomorphism of X onto $\hat{X}$. Consequently $\hat{X}$ is a Riesz subspace of X''. By making use of theorem (10.12)(b) and of the definition of normal integrals, we have $\hat{X} \subset (X')_n^b$, and hence $\hat{X} \subset X'' \cap (X')_n^b$. This completes the proof.

In the future, if no confusion can arise, we shall agree not to distinguish between x and $\hat{x}$; thus X will be identified with its canonical image $\hat{X}$.

The question naturally arises as to whether $\hat{X}$ is an ℓ-ideal in X''. In Chapter 13, we shall give necessary and sufficient condition to ensure that this occurs.

We conclude this section with a few examples of topological Riesz spaces.

(11.19) EXAMPLES. (a) Let T be a non-empty set. The vector space R^T of all real-valued functions on T is an order-complete Riesz space with the ordering induced by the cone

$$K = \{x \in R^T : x(t) \geqslant 0 \quad \text{for all} \quad t \in T\}.$$

Let X be a Riesz subspace of R^T, $\mathfrak{G}$ a family of subsets of T such that every x, in X, is bounded on each $B \in \mathfrak{G}$, and suppose that

$$T = \cup \{B : B \in \mathfrak{G}\}.$$

The topology on X of uniform convergence on members of $\mathfrak{G}$ is then a locally solid topology defined by the Riesz semi-norms $\{p_B : B \in \mathfrak{G}\}$, where $p_B(x) = \sup\{|x(t)| : t \in B\}$. If T is a Hausfdorff topological space, X is the Riesz space $C(T)$ consisting of all continuous real-valued functions on T, and if each $t \in T$ is interior to some $B \in \mathfrak{G}$, then X is complete by, for example, Bourbaki (1961). If T is a completely regular Hausdorff space and if $\mathfrak{G}$ is the family of all compact subsets of T, then the topology on $C(T)$ of uniform convergence on members of $\mathfrak{G}$ is called the *compact-open topology*, and denoted by $\mathscr{P}_c$. In particular, if T is a compact Hausdorff topological space then the compact-open topology on $C(T)$ is the uniform topology; in this case, $C(T)$ is a B-lattice.

(b) Let (X, C) be a Riesz space with the order-bound dual X^b, and let Y be a Riesz subspace of X^b such that Y is total over X. For any $f \in Y \cap C^*$, suppose that

$$p_f(x) = f(|x|) \qquad (x \in X).$$

Then the locally solid topology $|\sigma|(X, Y)$ determined by the family of Riesz semi-norms $\{p_f : f \in Y \cap C^*\}$ is generally called the *Dieudonné topology* (more precisely, the *Dieudonné topology induced by* Y). In next chapter we shall show that $|\sigma|(X, Y)$ is the locally solid topology associated with $\sigma(X, Y)$.

(c) Let $(T, \mathscr{B}, \mu)$ be a totally σ-finite measure space, and let $\{T_n : n \in \mathbf{N}\}$ be a fixed sequence in $\mathscr{B}$ such that $T = \bigcup_{n=1}^{\infty} T_n$ and $\mu(T_n) < \infty$ for all $n \in \mathbf{N}$. A real-valued function x on T is said to be *locally summable* if the restriction of x to each T_n is summable. Two functions on T are said to be *equivalent* if their difference is zero except on a set of μ-measure zero. Let Ω denote the set of all equivalent classes of locally summable functions on T. Ω has a natural cone C, defined by

$$C = \{x \in \Omega : x(t) \geqslant 0 \quad \text{for all} \quad t \in T\}.$$

It is easily seen that (Ω, C) is a Riesz space; in fact, (Ω, C) is order-complete. For a given set A in Ω, we define

$$\Lambda = A^{\times} = \{x \in \Omega : xy \in L^1(\mu) \quad \text{for all} \quad y \in A\},$$

$$\Lambda^{\times} = \{z \in \Omega : zx \in L^1(\mu) \quad \text{for all} \quad x \in \Lambda\},$$

then Λ and $\Lambda^{\times}$ are ℓ-ideals in Ω; Λ and $\Lambda^{\times}$ are placed in duality by the bilinear form

$$\langle x, z \rangle = \int_T xz \, d\mu \qquad x \in \Lambda \quad z \in \Lambda^{\times}.$$

Λ is called a *Köthe function space*, and $\Lambda^\times$ is referred to as the *Köthe dual* of Λ; of course, $\Lambda^\times$ is also a Köthe function space. It is not hard to show that Λ is the Köthe dual of $\Lambda^\times$, i.e. $\Lambda = \Lambda^{\times\times}$. Ω is the largest Köthe function space and $\Omega^\times$ is the smallest Köthe function space under the inclusion.

From now on, we always assume that Λ is a Köthe function space and that $\Lambda^\times$ is the Köthe dual of Λ. If $C_\Lambda = C \cap \Lambda$, then (Λ, C_Λ) is a Riesz space; moreover it is order-complete since Λ is an ℓ-ideal in Ω. If Λ^* is the algebraic dual of Λ and Λ^b is the order-bounded dual of (Λ, C_Λ), we then define a mapping ϕ of $\Lambda^\times$ into Λ^* by putting

$$\phi(z)(x) = \langle x, z \rangle \quad \text{for all} \quad x \in \Lambda,$$

where $z \in \Lambda^\times$; it is easily seen that ϕ is an ℓ-isomorphism of $\Lambda^\times$ into Λ^b, and hence $\phi(\Lambda^\times)$ is a Riesz subspace of Λ^b, consequently $\Lambda^\times$ can be regarded as a Riesz subspace of Λ^b.

According to the above remarks, $\sigma(\Lambda, \Lambda^\times)$ and $\sigma(\Lambda^\times, \Lambda)$ are locally o-convex topologies on Λ and $\Lambda^\times$ respectively, and $\Lambda^\times$ is the topological dual of $(\Lambda, \sigma(\Lambda, \Lambda^\times))$; Λ is the topological dual of $(\Lambda^\times, \sigma(\Lambda^\times, \Lambda))$. If $\sigma_S(\Lambda, \Lambda^\times)$ denotes the locally solid topology on Λ associated with $\sigma(\Lambda, \Lambda^\times)$ (for definition see Chapter 6), then $(\Lambda, C_\Lambda, \sigma_S(\Lambda, \Lambda^\times))$ is a locally convex Riesz space; in view of corollary (6.5), the topological dual $(\Lambda, C_\Lambda, \sigma_S(\Lambda, \Lambda^\times))'$ of $(\Lambda, C_\Lambda, \sigma_S(\Lambda, \Lambda^\times))$ is the ℓ-ideal in Λ^b generated by the Köthe dual $\Lambda^\times$ of Λ. Let $\mathfrak{G}$ be a family consisting of solid and $\sigma(\Lambda^\times, \Lambda)$-bounded sets in $\Lambda^\times$ which covers $\Lambda^\times$ and is directed by inclusion. Then there exists a unique locally solid topology, denoted by $\mathscr{P}_K$, on (Λ, C_Λ) such that $\{B^0 : B \in \mathfrak{G}\}$ is a neighbourhood-base at 0 for $\mathscr{P}_K$, so $(\Lambda, C_\Lambda, \mathscr{P}_K)$ is a locally convex Riesz space. This topology $\mathscr{P}_K$ is called a *Köthe topology*, and $(\Lambda, C_\Lambda, \mathscr{P}_K)$ is referred to as a *topological Köthe function space*. It should be noted that there is a finest Köthe topology, denoted by $\mathscr{P}_t$ on Λ obtained from $\Lambda^\times$ in this way; this is the one obtained by letting $\mathfrak{G}$ be the family of all solid and $\sigma(\Lambda^\times, \Lambda)$-bounded sets in $\Lambda^\times$. Also there is a coarsest Köthe topology on Λ obtained in the described way from $\Lambda^\times$; this is the one obtained by letting $\mathfrak{G}$ be the family of all order-intervals of the form $[-z, z]$ $(0 \leqslant z \in \Lambda^\times)$. We note that the coarsest Köthe topology on Λ is precisely the locally solid topology $\sigma_S(\Lambda, \Lambda^\times)$ associated with $\sigma(\Lambda, \Lambda^\times)$. Goffman (1959) has shown that, by virtue of the Nakano theorem, every topological Köthe function space is topologically complete.

(d) Let T be a locally compact Hausdorff space, and let $K(T)$ be the Riesz space of all real-valued and continuous functions on T with compact support, ordered by the positive cone C of non-negative functions in $K(T)$. For any compact subset L of T, let $K(T, L)$ be the set of all elements in $K(T)$ with support contained in L. It is easily seen that $K(T, L)$ is an ℓ-ideal in $K(T)$ and that $K(T)$ is the linear hull of $\{K(T, L): L$ is compact subset of $T\}$. For any compact subset L of T, the norm $\|\,.\,\|_L$, defined by

$$\|x\|_L = \sup\{|x(t)|: t \in L\} \qquad (x \in K(T, L)),$$

is a Riesz norm on $K(T, L)$. The inductive topology $\mathscr{T}_\mu$ on $K(T)$ with respect to the family of $\{(K(T, L), \|\,.\,\|_L): L$ is compact in $T\}$ and the injection maps $\{j_L: L$ is compact in $T\}$ is referred to as the *measure topology* on $K(T)$. In view of proposition (11.9), $\mathscr{T}_\mu$ is a locally solid topology on $K(T)$, and hence $(K(T), C, \mathscr{T}_\mu)$ is a (Hausdorff) locally convex Riesz space. Further we note that $(K(T), \mathscr{T}_\mu)$ is bornological (cf. Schaefer (1966)). The topological dual of $(K(T), C, \mathscr{T}_\mu)$ can be identified with the space $M(T)$ of all Radon measures on T (cf. R. E. Edwards (1965)), and hence from corollary (7.6) $M(T) = K(T)^b$ because $K(T)$ is $\mathscr{T}_\mu$-complete. Therefore each normal integral on $K(T)$ is a Radon measure on T; but the converse is not true as shown by Roberts (1964).

12

LOCALLY O-CONVEX RIESZ SPACES

FROM the preceding chapter we have seen that solid sets in a Riesz space play an important role in the study of topological Riesz spaces. Although locally o-convex topologies on a Riesz space are, in general, not locally solid (cf. example (3.15)), some results on locally convex Riesz spaces can be generalized to the case of locally o-convex Riesz spaces. On the other hand, we have seen from Chapter 6 that any locally o-convex topology $\mathscr{P}$ can be associated with a locally solid topology $\mathscr{P}_S$ which is the greatest lower bound of all locally solid topologies that are finer than $\mathscr{P}$. Now if $\mathscr{P}$ is a locally o-convex topology on a Riesz space (X, C) and if $\mathscr{U}$ is a neighbourhood-base at 0 for $\mathscr{P}$ consisting of o-convex circled sets in X, then

$$\mathscr{U}_S = \{\mathrm{sk}(V) : V \in \mathscr{U}\} = \{S(V) : V \in \mathscr{U}\}$$

is a neighbourhood-base at 0 for $\mathscr{P}_S$.

Let (X, C) be a Riesz space with the order-bound dual X^b, and let Y be a Riesz subspace of X^b which is total over X. Recall that the Dieudonné topology $|\sigma| (X, Y)$ on X, induced by Y, is defined to be the vector topology determined by the family $\{p_f : 0 \leqslant f \in Y\}$ of Riesz semi-norms, where

$$p_f(x) = f(|x|) \qquad (x \in X).$$

We now show that $|\sigma| (X, Y)$ is the locally solid topology associated with the weak topology $\sigma(X, Y)$.

(12.1) PROPOSITION. *Let (X, C) be a Riesz space with the order-bound dual X^b, and let Y be a Riesz subspace of X^b which is total over X. Then $|\sigma| (X, Y) = \sigma_S(X, Y)$ and the family*

$$\{[-f, f]^0 : 0 \leqslant f \in Y\}$$

forms a neighbourhood-base at 0 for $\sigma_S(X, Y)$. Furthermore, the topological dual of $(X, C, \sigma_S(X, Y))$ is the ℓ-ideal in X^b generated by Y. Therefore, if Y is an ℓ-ideal in X^b, then $\sigma_S(X, Y)$ is the coarsest locally solid topology which is consistent with the duality $\langle X, Y \rangle$.

Proof. Since Y is a Riesz subspace of X^b, then

$$Y = Y \cap C^* - Y \cap C^*;$$

in view of corollary (5.12), the weak topology $\sigma(X, Y)$ on X is locally o-convex. On the other hand, for any $0 \leqslant f \in Y$, if

$$V = \{x \in X : |f(x)| \leqslant 1\},$$

then $\text{sk}(V) = \{x \in X : [-|x|, |x|] \subseteq V\} = \{x \in X : f(|x|) \leqslant 1\}$, and it follows from proposition (10.15) that $\text{sk}(V) = [-f, f]^0$. Therefore $|\sigma| (X, Y) = \sigma_S(X, Y)$ and $\{[-f, f]^0 : 0 \leqslant f \in Y\}$ is a neighbourhood-base at 0 for $\sigma_S(X, Y)$. The result now follows from corollary (6.5) and the proof is complete.

(12.2) COROLLARY. *Let (X, C) be a Riesz with the order-bound dual X^b, and let Y be an ℓ-ideal in X^b which is total over X. Then the following statements hold:*

(a) *a subset M of X is $\sigma(X, Y)$-bounded if and only if its solid hull S_M is $\sigma(X, Y)$-bounded;*

(b) *each $\sigma(X, Y)$-bounded subset of X is uniformly bounded on each order-interval in Y;*

(c) *a subset B of Y is $\sigma_S(X, Y)$-equicontinuous if and only if it is an order-bounded subset of Y,*

(d) *a subset B of Y is $\sigma_S(Y, X)$-bounded if and only if S_B-$\sigma(Y, X)$-bounded.*

Proof. Since Y is an ℓ-ideal in X^b, then $\sigma_S(X, Y)$ is consistent with the duality $\langle X, Y \rangle$. The conclusion (a) follows from proposition (11.2) and the Mackey–Arens theorem, while the conclusion (b) is an immediate consequence of proposition (12.1). The conclusion (d) follows from the remark (c) of Theorem (10.12). It remains for us to verify (c). The sufficiency follows immediately from proposition (12.1); to prove the necessity, we note that the polar B^0 of B, taken in X, is a $\sigma_S(X, Y)$-neighbourhood of 0 in X, hence there exists $0 \leqslant f \in Y$ such that $[-f, f]^0 \subset B^0$, and hence $B \subset [-f, f]$. This completes the proof.

(12.3) COROLLARY. *Let (X, C) be a Riesz space and let Y be a Riesz subspace of X^b which is total over X. Then each positive $\sigma(Y, X)$-bounded subset B of Y is uniformly bounded on each order-interval in X.*

Proof. Elementary.

(12.4) PROPOSITION. *Let $\mathscr{P}$ be a locally o-convex topology on a Riesz space (X, C) and let $\mathscr{P}_S$ be the locally solid topology on X associated with $\mathscr{P}$. Then the following statements are equivalent:*
  (a) *C is a locally strict $\mathscr{B}$-cone in $(X, \mathscr{P})$;*
  (b) *each $\mathscr{P}$-bounded subset of X is $\mathscr{P}_S$-bounded;*
  (c) *C is a strict $\mathscr{B}$-cone in $(X, \mathscr{P})$.*

Proof. The implication (c) $\Rightarrow$ (a) is obvious, and the implication (a) $\Rightarrow$ (b) is a restatement of corollary (6.9). The observation that C is a strict $\mathscr{B}$-cone in $(X, \mathscr{P}_S)$ shows that (b) implies (c).

(12.5) COROLLARY. *Let $\mathscr{P}$ be a locally o-convex topology on a Riesz space (X, C). If the topological dual X' of $(X, C, \mathscr{P})$ is an ℓ-ideal in X^b then C is a strict $\mathscr{B}$-cone in $(X, \mathscr{P})$.*

Proof. Since X' is an ℓ-ideal in X^b, it follows from corollary (6.5) that the locally solid topology $\mathscr{P}_S$ associated with $\mathscr{P}$ is consistent with the duality $\langle X, X'\rangle$. The result now follows from proposition (12.4).

A partial converse of the preceding result is the following corollary.

(12.6) COROLLARY. *Let $\mathscr{P}$ be a locally o-convex topology on a Riesz space (X, C). Suppose that each circled o-convex set in X which absorbs all $\mathscr{P}$-bounded subsets of X is a $\mathscr{P}$-neighbourhood of 0 (i.e. $(X, C, \mathscr{P})$ is o-bornological in the sense of Kist (1958)). Then the following statements are equivalent:*
  (a) *$\mathscr{P}$ is locally solid;*
  (b) *the topological dual X' of $(X, C, \mathscr{P})$ is an ℓ-ideal in X^b;*
  (c) *C is a strict $\mathscr{B}$-cone in $(X, \mathscr{P})$;*
  (d) *C is a locally strict $\mathscr{B}$-cone in $(X, \mathscr{P})$.*

Proof. The implication (a) $\Rightarrow$ (b) follows from corollary (6.5), the implication (b) $\Rightarrow$ (c) and the equivalence of (c) and (d) follow from proposition (12.4). It remains to verify that (c) implies (a). Since $(X, C, \mathscr{P})$ is o-bornological, it is easily seen that no strictly finer locally o-convex topology on X has the same $\mathscr{P}$-bounded sets in X, it then follows from proposition (12.4) and from the definition of $\mathscr{P}_S$ that

$\mathscr{P} = \mathscr{P}_S$, and hence that $\mathscr{P}$ is locally solid. This completes the proof.

It should be noted from example (3.15) that the o-bornological condition in corollary (12.6) cannot be dropped. We shall see in Chapter 15 that under one of the assumptions (a), (b), (c), and (d) in the preceding result, $(X, C, \mathscr{P})$ is a bornological Riesz space.

By a *locally o-convex Riesz space* we mean a Riesz space (X, C) equipped with a locally o-convex topology $\mathscr{P}$ on X such that the cone C is $\mathscr{P}$-closed.

It is known from Peressini ((1961), theorem 2.4) that if $(X, C, \|\cdot\|)$ is a normed Riesz space with the topological dual X', then the weak topology $\sigma(X, X')$ is locally solid if and only if X is finite-dimensional. It is natural to ask under what conditions the Mackey topology $\tau(X, X')$ is locally solid. It will be shown that $\tau(X, X')$ is locally solid if and only if it is locally o-convex.

(12.7) PROPOSITION. *Let $(X, C, \mathscr{P})$ be a locally o-convex Riesz space with the topological dual X'. Then the Mackey topology $\tau(X, X')$ on X is locally solid if and only if $\tau(X, X')$ is a locally o-convex topology on X and X' is an ℓ-ideal in X^b.*

Proof. See theorem (6.1) and corollary (3.14).

(12.8) PROPOSITION. *Let $\{(X_\alpha, C_\alpha): \alpha \in \Gamma\}$ be a family of Riesz spaces, and let $\mathscr{P}_\alpha$ be a locally o-convex topology on (X_α, C_α) for each $\alpha \in \Gamma$. Suppose that $X = \prod_{\alpha \in \Gamma} X_\alpha$, $C = \prod_{\alpha \in \Gamma} C_\alpha$, and that $\mathscr{P} = \prod_{\alpha \in \Gamma} \mathscr{P}_\alpha$ (the product topology on X). Then $\mathscr{P}_S = \prod_{\alpha \in \Gamma} \mathscr{P}_{\alpha, S}$, and therefore the ℓ-ideal in X^b generated by $\bigoplus_{\alpha \in \Gamma} (X_\alpha, C_\alpha, \mathscr{P}_\alpha)'$ is precisely the algebraic direct sum of ℓ-ideals in X_α^b generated by $(X_\alpha, C_\alpha, \mathscr{P}_\alpha)'$ ($\alpha \in \Gamma$), where $\mathscr{P}_{\alpha, S}$ denotes the locally solid topology on X_α associated with $\mathscr{P}_\alpha$ for each $\alpha \in \Gamma$ and $\mathscr{P}_S$ denotes the locally solid topology on X associated with $\mathscr{P}$.*

Proof. See theorems (3.19) and (6.1).

(12.9) PROPOSITION. *Let $\{(X_\alpha, C_\alpha): \alpha \in \Gamma\}$ be a family of Riesz spaces, and let $\mathscr{P}_\alpha$ be a locally o-convex topology on (X_α, C_α) for each $\alpha \in \Gamma$. Suppose that $Y := \bigoplus_{\alpha \in \Gamma} X_\alpha$ (the algebraic direct sum of $\{X_\alpha: \alpha \in \Gamma\}$), $K = Y \cap \prod_\alpha C_\alpha$, and that $\mathscr{P}' = \bigoplus_\alpha \mathscr{P}_\alpha$ (the locally convex sum topology of $\mathscr{P}_\alpha$ on Y). Then $\mathscr{P}'_S = \bigoplus_\alpha \mathscr{P}_{\alpha, S}$, and therefore the ℓ-ideal* *in* Y^b *generated by* $\prod_{\alpha \in \Gamma} (X_\alpha, C_\alpha, \mathscr{P}_\alpha)'$ *is precisely the product of ℓ-ideals in* X_α^b *generated by* $(X_\alpha, C_\alpha, \mathscr{P}_\alpha)'$ $(\alpha \in \Gamma)$, *where* $\mathscr{P}_{\alpha,\mathrm{S}}$ *denotes the locally solid topology on X_α associated with* $\mathscr{P}_\alpha$ *for each* $\alpha \in \Gamma$, *and* $\mathscr{P}'_\mathrm{S}$ *denotes the locally solid topology on Y associated with* $\mathscr{P}'$.

Proof. See theorems (3.16) and (6.1).

We conclude this chapter with a result concerning the locally order-completeness of $(X', C', \sigma_\mathrm{S}(X', X))$.

(12.10) PROPOSITION. *Let* $(X, C, \mathscr{P})$ *be a locally o-convex Riesz space with the topological dual X'. If X' is an ℓ-ideal in X^b, then* $(X', C', \sigma_\mathrm{S}(X', X))$ *is always locally order-complete.*

Proof. Since X' is an ℓ-ideal in X^b, it follows that $(X', C', \sigma_\mathrm{S}(X', X))$ is a locally convex Riesz space. In view of proposition (12.1), $\{[-u, u]^0 : u \in C\}$ is a neighbourhood-base at 0 for $\sigma_\mathrm{S}(X', X)$. In order to verify this result, it is sufficient to show that each $[-u, u]^0$ is order-complete. Let $\{f_\tau\}$ be an increasing net in $[-u, u]^0$ which is majorized in X'. Then $f = \sup f_\tau$ exists in X' because (X', C') is order-complete. It is clear that $f(w) = \sup f_\tau(w)$ for any $w \in C$; then f_τ converges to f with respect to $\sigma(X', X)$, and hence, from the $\sigma(X', X)$-closedness of $[-u, u]^0$, we have $f \in [-u, u]^0$. Thus each $[-u, u]^0$ is order-complete, consequently $(X', C', \sigma_\mathrm{S}(X', X))$ is locally order-complete.

13

COMPLETENESS FOR THE DIEUDONNÉ TOPOLOGY

I T is known from proposition (11.18) that a locally convex Riesz space X can be embedded as a Riesz subspace of $X'' \cap (X')_n^b$. The following question naturally arises:

(1) What condition is necessary and sufficient for the embedding to preserve the supremum and infimum for infinite subsets of X?

This also suggests the following two intimately related questions:

(2) What condition on X (or X') is necessary and sufficient for X to be an ℓ-ideal in $X'' \cap (X')_n^b$?

(3) What condition on X (or X') is necessary and sufficient for X to be a normal subspace of $(X')_n^b$?

This chapter is devoted to answering these and other related questions. We shall see in particular that the answer to (3) relies on the converse of the Nakano theorem or, equivalently, on the completeness property of the Dieudonné topology (see theorem (13.9)).

(13.1) THEOREM (Andô–Luxemburg–Zaanen). *Let $(X, C, \mathscr{P})$ be a locally convex Riesz space with the topological dual X'. Then the following statements are equivalent:*

(a) $X' \subseteq X_n^b$;

(b) *if $w_\tau \downarrow 0$ ($\tau \in D$) then w_τ converges to 0 with respect to $\mathscr{P}$;*

(c) *the ℓ-ideal in X'' generated by X is an order completion of X;*

(d) *the supremum and infimum of any subset of X are preserved under the evaluation map $x \to \hat{x}$ of X into $X'' \cap (X')_n^b$;*

(e) *for any $f \in X'$, the ℓ-ideal $N_f = \{x \in X : |f|(|x|) = 0\}$ is a normal subspace of X;*

(f) *each order-dense ℓ-ideal in X is $\mathscr{P}$-dense in X;*

(g) *each $\mathscr{P}$-closed ℓ-ideal in X is a normal subspace of X.*

Proof. In view of proposition (5.8), it is clear that (a) $\Leftrightarrow$ (b).

(a) $\Rightarrow$ (c): Let L be the ℓ-ideal in $X'' \cap (X')_n^b$ generated by X. In order to verify that L is an order-completion of X, it is sufficient to verify that, for each $0 < u'' \in L$, there exist $u_0, u \in X$ such that

$0 < \hat{u}_0 \leqslant u'' \leqslant \hat{u}$. Since X is a Riesz subspace of $X'' \cap (X')_n^b$, the generated ℓ-ideal L must be the order-convex hull of X; hence $u'' \leqslant \hat{u}$ for some $u \in X$. Since $u'' > 0$, for this u, we can take a positive real number λ small enough such that $(u'' - \lambda \hat{u})^+ > 0$. For such a λ let $v_\lambda = (u'' - \lambda \hat{u})$. Then, $v_\lambda, v_\lambda^+, v_\lambda^-$ are all in $X'' \cap (X')_n^b$. Let

$$A = \{\psi \in X' : v_\lambda^+(\psi) = 0\}.$$

Since v_λ^+ is a normal integral, A is a normal subspace of X'; also $X' = A \oplus A^d$ by corollary (10.11). Since $v_\lambda^+ > 0$, it follows that $X' \neq A$ and hence there exists $0 < h \in A^d$ such that $v_\lambda^+(h) > 0$. Since $v_\lambda^+ \leqslant u'' \leqslant \hat{u}$, $h(u) = \hat{u}(h) \geqslant v_\lambda^+(h) > 0$ for this h. Since (a) holds, apply proposition (10.20) to obtain an element $w \in X$ with $0 < w \leqslant u$ such that

$$h(w) > 0 \quad \text{and} \quad \psi(w) = 0 \quad \text{for all } \psi \in X_n^b \text{ with } \psi \perp h;$$

thus, in particular, $\psi(w) = 0$ for all $\psi \in A$.

Then $\lambda w \in X$ and $0 < \lambda \hat{w} \leqslant u''$. To verify this, let $0 \leqslant \psi \in X'$, and suppose that $\psi = \psi_1 + \psi_2$, where $0 \leqslant \psi_1 \in A$ and $0 \leqslant \psi_2 \in A''$. Then

$$(\lambda \hat{w})(\psi_1) = \lambda \psi_1(w) = 0 \leqslant u''(\psi_1).$$

Also, since $v_\lambda^- \perp v_\lambda^+$, we recall from corollary (10.19) that v_λ^- must vanish on A^d, and, in particular, that $v_\lambda^-(\psi_2) = 0$. Consequently

$$(u'' - \lambda \hat{w})(\psi_2) \geqslant (u'' - \lambda \hat{u})(\psi_2) = v_\lambda(\psi_2)$$
$$= v_\lambda^+(\psi_2) - v_\lambda^-(\psi_2) = v_\lambda^+(\psi_2) \geqslant 0.$$

This, together with an earlier established inequality, implies that

$$(u'' - \lambda \hat{w})(\psi) = (u'' - \lambda \hat{w})(\psi_1) + (u'' - \lambda \hat{w})(\psi_2) \geqslant 0,$$

therefore $u'' \geqslant \lambda \hat{w}$, as claimed.

(c) $\Rightarrow$ (d): Let $\{x_\tau : \tau \in D\}$ be a subset of X and let x be the supremum in X of the set $\{x_\tau\}$. We assume without loss of generality that $0 \in \{x_\tau\}$. Let L be the ℓ-ideal in X'' generated by X. Then x is also the supremum in L of $\{x_\tau\}$ since L is an order-completion of X. Finally, since L is an ℓ-ideal in X'', x must in fact be the supremum of $\{x_\tau\}$ in X''. To verify this, let y be an element of X'' majorizing $\{x_\tau\}$. Then $0 \leqslant y \wedge x \leqslant x \in L$, so $y \wedge x \in L$ and is an upper bound of $\{x_\tau\}$ in L, hence $x \leqslant x \wedge y$ since x is the supremum of $\{x_\tau\}$ in L. Therefore $x \leqslant y$, and this implies that x is the supremum of $\{x_\tau\}$ in X''.

(d) $\Rightarrow$ (e): Let $f \in X'$ and let $0 \leqslant u_\tau \uparrow u$ in X, where $u_\tau \in N_f$. Then, by (d), we have $0 \leqslant \hat{u}_\tau \uparrow \hat{u}$ in X'', i.e.

$$\sup \hat{u}_\tau(\psi) = \hat{u}(\psi) \qquad (\psi \in X', \psi \geqslant 0).$$

In particular, since $u_\tau \in N_f$,

$$0 = \sup |f|(u_\tau) = \sup \hat{u}_\tau(|f|) = \hat{u}(|f|) = |f|(u),$$

showing that $u \in N_f$. Therefore N_f is a normal subspace of X.

(e) $\Rightarrow$ (f): Let A be an order-dense ℓ-ideal in X. If A is not $\mathscr{P}$-dense in X, then there exists $u \in C$ such that $u \notin \bar{A}$. By the separation theorem, there exists a $g \in X'$ such that

$$g(u) \neq 0 \quad \text{and} \quad g(a) = 0 \quad \text{for all } a \text{ in } A.$$

Let $f = |g|$. Then $f \in X'$,

$$f(u) = \sup\{|g(x)| : |x| \leqslant u\} \geqslant |g(u)| > 0,$$

and

$$f(a) = 0 \quad \text{for all } a \text{ in } A.$$

Since A is an ℓ-ideal, it follows that $A \subset N_f$. By (e), we then have $\{A\} \subset N_f$. However, since A is order-dense in X, $\{A\} = X$, that is, $X = N_f$, contrary to the fact that $f(u) \neq 0$. Therefore A must be $\mathscr{P}$-dense in X.

(f) $\Rightarrow$ (g): Let B be a $\mathscr{P}$-closed ℓ-ideal in X, and let $\{B\}$ be the normal subspace in X generated by B. Since X is Archimedean, by proposition (10.9), we have that $\{B\} = B^{dd}$ and that $B \oplus B^d$ is order-dense in X. By (f) it follows that $B \oplus B^d$ is $\mathscr{P}$-dense in X. In particular, if $0 < u \in \{B\}$, then there exists a net $\{u_\tau : \tau \in D\}$ in $B \oplus B^d$ such that u_τ converges to u with respect to $\mathscr{P}$. For each $\tau \in D$, let $w_\tau = u_\tau^+ \wedge u$. Then $0 \leqslant w_\tau \leqslant u$, $w_\tau \in B \oplus B^d$, and w_τ converges to u with respect to $\mathscr{P}$ because the lattice operations are $\mathscr{P}$-continuous. Since $0 \leqslant w_\tau \leqslant u \in \{B\} = B^{dd}$, each $w_\tau \in B^{dd}$. We shall show that $w_\tau \in B$. In fact, write

$$w_\tau = w_\tau' + w_\tau'' \in B \oplus B^d,$$

where $w_\tau' \in B$ and $w_\tau'' \in B^d$. Then, since $B \subset B^{dd}$, $w_\tau' \in B^{dd}$,

$$w_\tau'' = w_\tau - w_\tau' \in B^{dd} - B^{dd} \subset B^{dd}.$$

But we also have $w_\tau'' \in B^d$; it follows that $w_\tau'' = 0$. Hence $w_\tau = w_\tau' \in B$. Since u is the $\mathscr{P}$-limit of $\{w_\tau\}$, it follows that $u \in \bar{B}$. Since B is $\mathscr{P}$-closed, $u \in B$. This shows that $\{B\} \subset B$. Consequently $\{B\} = B$ and B is a normal subspace of X.

Finally we show that (g) $\Rightarrow$ (b). Clearly (b) is equivalent to the following statement:

(b') If $0 \leqslant u_\tau \uparrow u$ then u_τ converges to u with respect to $\mathscr{P}$.

Suppose that $0 \leqslant u_\tau \uparrow u$, and let W be a $\mathscr{P}$-neighbourhood of 0 in X. Take a convex and solid $\mathscr{P}$-neighbourhood V of 0 such that $V + V \subset W$. Choose a real number α with $0 < \alpha < 1$ such that $(1-\alpha)u \in V$. For each τ in the index set D, let

$$v_\tau = u_\tau - \alpha u$$

and let A be the ℓ-ideal in X generated by $\{v_\tau : \tau \in D\}$. Since $v_\tau \uparrow u - \alpha u$, it is clear that $u - \alpha u \in \{A\}$, hence $u \in \{A\}$. On the other hand, let $\bar{A}$ denote the $\mathscr{P}$-closed of A. Then $\bar{A}$ is also an ℓ-ideal and hence must be a normal subspace in X by (g). Therefore $\bar{A} = \{A\}$ and $u \in \bar{A}$. Take an element w in A such that $u - w \in V$. Then $w^+ \in A$ and hence

$$0 \leqslant w^+ \leqslant nv_{\tau_0}$$

for some positive integer n and some $\tau_0 \in D$. Thus $0 \leqslant w^+ \leqslant nv_{\tau_0}^+$; and since $v_{\tau_0}^-$ is disjoint from $v_{\tau_0}^+$, it must be disjoint from w^+ and $w^+ \wedge u$. Hence

$$v_{\tau_0}^- + w^+ \wedge u = v_{\tau_0}^- \vee (w^+ \wedge u) \leqslant u$$

and

$$0 \leqslant v_{\tau_0}^- \leqslant u - w^+ \wedge u \leqslant (u - w^+) \vee 0 \leqslant |u^+ - w^+| \leqslant |u - w| \in V.$$

Since V is solid, we see that $v_{\tau_0}^- \in V$. Now, for all $\tau \geqslant \tau_0$ in D, we have

$$
\begin{aligned}
u - u_\tau &\leqslant u - u_{\tau_0} = u - u_{\tau_0} - \alpha u + \alpha u \\
&= (1-\alpha)u + (u_{\tau_0} - \alpha u)^- - (u_{\tau_0} - \alpha u)^+ \leqslant (1-\alpha)u + (u_{\tau_0} - \alpha u)^- \\
&= (1-\alpha)u + v_{\tau_0}^- \in V + V \subset W.
\end{aligned}
$$

This implies that u_τ converges to u with respect to $\mathscr{P}$; thus (b') is true and hence (b) is proved.

(13.2) COROLLARY. *For any Riesz space* (X, C), *the following statements are equivalent*:

(a) $X_c^b = X_n^b$;

(b) *for each* $f \in X_c^b$, *the* ℓ-*ideal* $N_f = \{x \in X : |f|\,(|x|) = 0\}$ *is a normal subspace of* X;

(c) *each* $\sigma(X, X_c^b)$-*closed* ℓ-*ideal in* X *is a normal subspace of* X.

Proof. It is clear that $\sigma(X, X_c^b)$ is a locally o-convex topology on X. Let $\sigma_S(X, X_c^b)$ be the locally solid topology on X associated with $\sigma(X, X_c^b)$. Since X_c^b is an ℓ-ideal in X^b, it follows from corollary (6.5) that $X_c^b = (X, C, \sigma_S(X, X_c^b))'$ and hence that an ℓ-ideal in X is $\sigma_S(X, X_c^b)$-closed if and only if it is $\sigma(X, X_c^b)$-closed. The result now follows immediately from the preceding theorem.

(13.3) COROLLARY. *Let $(X, C, \mathscr{P})$ be a locally o-convex Riesz space with the topological dual X'. Then $X' \subset X_n^b$ if and only if u_τ converges to 0 with respect to $\mathscr{P}$ whenever $u_\tau \downarrow 0$.*

Proof. Let $\mathscr{P}_S$ be the locally solid topology on X associated with $\mathscr{P}$. Then $(X, C, \mathscr{P}_S)' \subset X_n^b$ if and only if u_τ converges to 0 with respect to $\mathscr{P}_S$ whenever $u_\tau \downarrow 0$, in view of theorem (13.1). Therefore the result follows from (c) of corollary (6.9) and corollary (6.5). We note that this corollary can also be proved directly from proposition (5.8).

The proof of the following proposition is similar to that given in theorem (13.1), and will be omitted.

(13.4) PROPOSITION. *Let $(X, C, \mathscr{P})$ be a locally convex Riesz space. Then the following statements are equivalent:*

(a) $X' \subset X_c^b$;

(b) *if $u_n \downarrow 0$ $(n \in \mathbf{N})$ then u_n converges to 0 with respect to $\mathscr{P}$;*

(c) *the supremum and infimum of any countable subset of X are preserved under the evaluation map $x \to \hat{x}$ of X into $X'' \cap (X')_n^b$.*

We are now going to seek some necessary and sufficient condition ensuring that X can be regarded as an ℓ-ideal in X''. Parts of the following theorem, namely the equivalence of (a), (b), and (d), were proved by Kawai (1957).

(13.5) THEOREM. *Let $(X, C, \mathscr{P})$ be a locally convex Riesz space with the topological dual X'. Then the following statements are equivalent:*

(a) $X' \subset X_n^b$ *and (X, C) is order-complete;*

(b) *X is ℓ-isomorphic with an ℓ-ideal in X'' under the evaluation map $x \to \hat{x}$ of X into $X'' \cap (X')_n^b$;*

(c) *the Dieudonné topology $\sigma_S(X', X)$ on X' is consistent with the duality $\langle X, X' \rangle$;*

(d) *each order-interval in X is $\sigma(X, X')$-compact;*

(e) *each order-bounded subset of X which is directed upwards has a $\mathscr{P}$-limit.*

Furthermore, if $(X, C, \mathscr{P})$ satisfies one (and hence all) of (a)–(e) then $(X, C, \mathscr{P})$ is locally order-complete.

Proof. Recall from corollary (6.5) that the topological dual of $(X', \sigma_S(X', X))$ is the ℓ-ideal in X'' generated by $X = \hat{X}$; thus the equivalence of (b) and (c) is clear. Proposition (12.1) tells us that $\sigma_S(X', X)$ is the topology of uniform convergence on order-intervals in X, hence the equivalence of (c) and (d) is just a restatement of the Makey–Arens theorem. Further, by proposition (5.8), a directed upwards net in X converges with respect to $\mathscr{P}$ if and only if it does with respect to $\sigma(X, X')$; hence (d) implies (e). Therefore, to complete the proof, it remains to verify that (a) $\Rightarrow$ (b) and (e) $\Rightarrow$ (a).

(a) $\Rightarrow$ (b): Suppose (a) holds. In view of theorem (13.1), the ℓ-ideal L in X'' generated by X is an order-completion of X. However, since X is already order-complete, it follows that $X = L$.

(e) $\Rightarrow$ (a): The order completeness of X is obvious since C is $\mathscr{P}$-closed. On the other hand, if $0 \leqslant u_\tau \downarrow 0$ in X, for any fixed τ_0, let $w_\tau = \inf\{u_\tau, u_{\tau_0}\}$. Then $0 \leqslant w_\tau \leqslant u_{\tau_0}$ $0 \leqslant w_\tau \leqslant u_\tau$, so that $w_\tau \downarrow 0$, we then have that $0 \leqslant u_{\tau_0} - w_\tau \uparrow u_{\tau_0}$ in X. In view of the assumption (e), there exists $u \in X$ such that $u_{\tau_0} - w_\tau$ converges to u with respect to $\mathscr{P}$. It follows from proposition (2.1) that $u_{\tau_0} - w_\tau \uparrow u$, and hence from $u_{\tau_0} - w_\tau \uparrow u_{\tau_0}$ that $u = u_{\tau_0}$. Therefore w_τ converges to 0 with respect to $\mathscr{P}$. Since u_τ is decreasing, then u_τ converges to 0 with respect to $\mathscr{P}$, and so, by making use of theorem (13.1), $X' \subset X_n^b$.

Finally, let $\mathscr{U}$ be a neighbourhood-base at 0 in $(X, \mathscr{P})$ consisting of $\mathscr{P}$-closed solid sets. Then each number V in $\mathscr{U}$ is order-complete. In fact, let x_τ in V be such that $x_\tau \uparrow \leqslant y$ for some $y \in X$. Since X is order-complete, there exists $x \in X$ such that $x = \sup x_\tau$. It follows from $X' \subset X_n^b$ that x_τ converges to x with respect to $\mathscr{P}$, and hence that $x \in V$. Therefore each V is an order-complete set in X, consequently $(X, C, \mathscr{P})$ is locally order-complete. This completes the entire proof of this theorem.

(13.6) COROLLARY. *Let $(X, C, \mathscr{P})$ be a locally o-convex Riesz space with the topological dual X'. Then the following statements are equivalent:*

(a) $X' \subset X_n^b$ *and (X, C) is order-complete;*

(b) *each order-bounded subset of X which is directed upwards has a $\mathscr{P}$-limit.*

Furthermore, if $\mathscr{P}_S$ denotes the locally solid topology associated with $\mathscr{P}$, and if $(X, C, \mathscr{P})$ satisfies one of (a) and (b), then $(X, C, \mathscr{P}_S)$ is locally order-complete.

Proof. It is known from corollary (6.5) that $X' \subset X_n^b$ if and only if $(X, C, \mathscr{P}_S)' \subset X_n^b$. From corollary (6.9), u_τ converges to u with respect to $\mathscr{P}$ if and only if u_τ converges to u with respect to $\mathscr{P}_S$ whenever

$u_\tau \uparrow$. Therefore the result follows immediately from the preceding theorem.

(13.7) COROLLARY. *Let (X, C) be a Riesz space, X_n^b total over X, and suppose that X_{nn} is the set of all normal integrals on X_n^b, that is $X_{nn} = (X_n^b)_n^b$. Then the following statements are equivalent:*

(a) *X is order-complete;*

(b) *X is ℓ-isomorphic with an ℓ-ideal in X_{nn} under the evaluation map $x \to \hat{x}$ defined by $\hat{x}(f) = f(x)$ for all $f \in X_n^b$.*

Furthermore, if X satisfies one of (a) and (b) then the smallest normal subspace of X_{nn} generated by X is precisely X_{nn}.

Proof. Since X_n^b is a normal subspace of X^b and certainly an ℓ-ideal, $(X, C, \sigma_S(X, X_n^b))$ is a locally convex Riesz space with the topological dual X_n^b, and so (b) implies (a) in view of theorem (13.5). Conversely, if X is order-complete, then theorem (13.5) shows that X is an ℓ-ideal in $(X_n^b, \beta(X_n^b, X))'$ and hence in $(X_n^b)^b$. Note also that $X \subset X_{nn}$ and that X_{nn} is an ℓ-ideal in $(X_n^b)^b$; then X is an ℓ-ideal in X_{nn}, and consequently (a) implies (b). Finally, since X_n^b is order-complete and since X is an ℓ-ideal in X_{nn}, it follows from corollary (10.22) that $\{X\} = (X^0)_n^\pi$, where $X^0 = \{f \in X_n^b : \hat{x}(f) = f(x) = 0 \; \forall \, x \in X\}$. Since X_n^b is total over X then $X^0 = \{0\}$ and so $(X^0)_n^\pi = X_{nn}$; consequently the smallest normal subspace of X_{nn} generated by X is exactly X_{nn}.

(13.8) COROLLARY. *Let (X, C) be an order-complete Riesz space, and let X_n^b be total over X. If $\tau(X, X_n^b)$ denotes the Mackey topology on X with respect to the dual pair $\langle X, X_n^b \rangle$, then the following statements are equivalent and each is equivalent to each of (a)–(g) of theorem (13.1):*

(a) *$(X, C, \tau(X, X_n^b))$ is barrelled;*

(b) *each $\beta(X, X_n^b)$-closed ℓ-ideal in X is a normal subspace of X_n^b;*

(c) *if $u_\tau \uparrow u$ in X then u_τ converges uniformly to u on each $\sigma(X_n^b, X)$-bounded subset of X_n^b.*

Proof. In view of corollary (13.7), X is an ℓ-ideal in X_{nn}, then $(X_n^b, \sigma_S(X_n^b, X))' = X$, and hence the solid hull of each $\sigma(X_n^b, X)$-bounded subset of X_n^b is $\sigma(X_n^b, X)$-bounded because the topologies $\sigma(X_n^b, X)$ and $\sigma_S(X_n^b, X)$ are consistent with the dual pair $\langle X, X_n^b \rangle$. Consequently $\beta(X, X_n^b)$ is a locally solid topology on X. Now suppose that $(X, C, \tau(X, X_n^b))$ is barrelled. Then each $\sigma(X_n^b, X)$-bounded subset of X_n^b is $\tau(X, X_n^b)$-equicontinuous, and hence $\tau(X, X_n^b)$ is exactly $\beta(X, X_n^b)$; consequently $(X, C, \beta(X, X_n^b))' = X_n^b$. By theorem (13.1), each $\beta(X, X_n^b)$-closed ℓ-ideal in X is a normal subspace of X, thus (a) implies (b). If $u_\tau \uparrow u$ in X and if the statement (b) holds then, in view of theorem (13.1), u_τ converges to u with respect to $\beta(X, X_n^b)$; in other words, u_τ converges uniformly to u on each $\sigma(X_n^b, X)$-bounded subset of X_n^b. Finally suppose that the statement (c) holds. Then w_τ converges to 0 with respect to $\beta(X, X_n^b)$ provided that $w_\tau \downarrow 0$; hence, according to theorem (13.1), $(X, C, \beta(X, X_n^b))' \subseteq X_n^b$, and consequently $\beta(X, X_n^b)$ is consistent with the dual pair $\langle X, X_n^b \rangle$. Therefore $\beta(X, X_n^b)$ is exactly $\tau(X, X_n^b)$, and so $(X, C, \tau(X, X_n^b))$ is barrelled. This completes the proof.

We are now in a position to deal with the final question posed at the beginning of this chapter, namely: What condition on X (or X') is necessary and sufficient to ensure that X is a normal subspace of $(X')_n^b$? We shall see that the answer to this question is equivalent to the completeness of X for the Dieudonné topology. On the other hand, it is known from example (11.15) that, in general, the converse of Nakano's theorem (11.14) fails; therefore it is interesting to find some classes of locally convex Riesz spaces for which the converse of Nakano's theorem holds. The following result of Wong (1969c) shows that the completeness of X for the Dieudonné topology ensures that the converse of Nakano's theorem holds.

(13.9) THEOREM. *Let* $(X, C, \mathscr{P})$ *be a locally* o-*convex Riesz space with the topological dual* X'. *Then the following statements are equivalent:*

(a) $X' \subset X_n^b$ *and* $(X, C, \mathscr{P})$ *is boundedly order-complete;*

(b) $(X, C, \sigma_S(X, X'))$ *is both locally order-complete and boundedly order-complete;*

(c) X *is complete for* $\sigma_S(X, X')$;

(d) *each positive* $\sigma(X, X')$-*bounded subset of* X *which is directed upwards has a* $\sigma(X, X')$-*limit;*

(e) *each positive* $\mathscr{P}$-*bounded subset of* X *which is directed upwards has a* $\mathscr{P}$-*limit.*

Furthermore, if $(X, C, \mathscr{P})$ *satisfies one of* (a)–(e) *then* X *is complete for* $\mathscr{P}_S$.

Proof. (a) $\Rightarrow$ (b): Since $\sigma(X, X')$ and $\mathscr{P}$ are consistent with the duality $\langle X, X' \rangle$, then the locally o-convex Riesz space $(X, C, \sigma(X, X'))$ is boundedly order-complete; in particular, $(X, C, \sigma_S(X, X'))$ is boundedly order-complete and X is order-complete. According to corollary (13.6), $(X, C, \sigma_S(X, X'))$ is locally order-complete.

(b) $\Rightarrow$ (c): Follows from Nakano's theorem (11.14).

(c) $\Rightarrow$ (d): Let $\{u_\tau\}$ be a positive $\sigma(X, X')$-bounded subset of X which is directed upwards. Then, for any $f \in C'$, $\{f(u_\tau)\}$ is a bounded increasing net of real numbers. It follows that $\{u_\tau\}$ is a $\sigma(X, X')$-Cauchy net and consequently a $\sigma_S(X, X')$-Cauchy net since $u_\tau\uparrow$. There exists $u \in X$ such that u_τ converges to u with respect to $\sigma_S(X, X')$ and *a fortiori* with respect to $\sigma(X, X')$.

(d) $\Rightarrow$ (e): Let $\{u_\tau\}$ be a positive $\mathscr{P}$-bounded subset of X which is directed upwards. The $\{u_\tau\}$ is $\sigma(X, X')$-bounded, and so there exists $u \in X$ such that u_τ converges to u with respect to $\sigma(X, X')$; we conclude from proposition (5.8) that u_τ converges to u with respect to $\mathscr{P}$.

(e) $\Rightarrow$ (a): It is clear that $(X, C, \mathscr{P})$ is boundedly order-complete. On the other hand, suppose $u_\tau \downarrow 0$. Without loss of generality we can assume that $u_\tau \leqslant u$ for some $u \in C$. Then $\{u-u_\tau\}$ is a positive order-bounded subset of X which is directed upwards, and so $\{u-u_\tau\}$ is $\mathscr{P}$-bounded. There exists $u_0 \in X$ such that $u-u_\tau$ converges to u_0 with respect to $\mathscr{P}$ so, by the closedness of C, $u-u_\tau \uparrow u_0$, and hence $u = u_0$ because $u-u_\tau \uparrow u$; consequently u_τ converges to 0 with respect to $\mathscr{P}$. In view of corollary (13.3), $X' \subset X_n^b$.

To see the final assertion of the theorem, it suffices to remark that $\mathscr{P}_S \geqslant \sigma_S(X, X')$ and both topologies admit the same topological dual.

The following corollary is a dual result of theorem (13.9).

(13.10) COROLLARY.   *Let $(X, C, \mathscr{P})$ be a locally convex Riesz space with the topological dual X'. Then the following statements are equivalent:*

(a) *$(X', C', \sigma(X', X))$ is boundedly order-complete;*

(b) *$(X', C', \sigma_S(X', X))$ is both locally order-complete and boundedly order-complete;*

(c) *X' is complete for $\sigma_S(X', X)$;*

(d) *each positive $\sigma(X', X)$-bounded subset of X' which is directed upwards has a $\sigma(X', X)$-limit;*

(e) *X' is a normal subspace of X^b.*

Proof.   It should be noted that $(X', C', \sigma(X', X))$ is a locally o-convex Riesz space for which the topological dual of $(X', C', \sigma(X', X))$ is contained in $(X')_n^b$, and then the equivalence of (a), (b), (c), and (d) follows immediately from theorem (13.9). It remains to verify the implications (d) $\Rightarrow$ (e) $\Rightarrow$ (a).

(d) $\Rightarrow$ (e): Let $\{f_\tau\}$ be a positive and directed upwards subset of X', and let f in X^b be such that $f_\tau \uparrow f$. The $f(u) = \sup f_\tau(u)$ for any $u \in C$, and so $\{f_\tau\}$ is $\sigma(X', X)$-bounded. By the statement (d), there exists $g \in X'$ such that f_τ converges to g with respect to $\sigma(X', X)$, so $f_\tau \uparrow g$

since C' is $\sigma(X', X)$-closed, hence $f = g$ and $f \in X'$; consequently X' is a normal subspace of X^b.

(e) $\Rightarrow$ (a): Let $\{f_\tau : \tau \in D\}$ be a $\sigma(X', X)$-bounded and monotonically increasing net in X'. Then, for each $u \in C$, $\{f_\tau(u) : \tau \in D\}$ is a bounded and increasing net of real numbers; hence the net converges to a real number, denoted by $f(u)$. Then $u \to f(u)$ is a well-defined, additive, and positively homogeneous functional on C. This function f can be extended to be defined on the whole X. Then $f \in X^b$ and $f_\tau \uparrow f$, and it follows from (e) that $f \in X'$. Thus (a) holds.

The completion of $(X', \sigma_S(X', X))$ coincides exactly with the normal subspace of X^b generated by X', as the following result (due to Peressini (1967)) shows.

(13.11) COROLLARY. *Let* $(X, C, \mathscr{P})$ *be a locally convex Riesz space with the topological dual* X'. *Then the completion of* X' *for* $\sigma_S(X', X)$ *is the normal subspace of* X^b *generated by* X'.

Proof. Let Y be the normal subspace of X^b generated by X'. Then $\sigma_S(X, Y)$ is a locally solid topology on X such that Y is the topological dual of $(X, C, \sigma_S(X, Y))$, so, by corollary (13.10), Y is complete for $\sigma_S(Y, X)$. It is clear that for each $x \in X$, the linear functional $\hat{x}$, defined by
$$\hat{x}(f) = f(x) \quad \text{for all } f \in Y,$$
is a normal integral on Y; hence $(Y, \sigma(Y, X))' \subset Y_n^b$ and consequently $(Y, \sigma_S(Y, X))' \subset Y_n^b$. Furthermore, X' is an ℓ-ideal in Y, it follows from theorem (13.1) that X' is dense in Y with respect to $\sigma_S(Y, X)$. Since the topology $\sigma_S(X', X)$ on X' is the relative topology induced by $\sigma_S(Y, X)$, we conclude that Y is the completion of X' for $\sigma_S(X', X)$.

(13.12) COROLLARY. *Let* $(X, C, \mathscr{P})$ *be a locally convex Riesz space with the topological dual* X'. *Then the following statements are equivalent and each is equivalent to each of* (a), (b), (d), *and* (e) *of theorem* (13.9):
   (a) X *is a normal subspace of* $(X')_n^b$;
   (b) $X = (X')_n^b$;
   (c) X *is complete for* $\sigma_S(X, X')$.

Proof. (a) $\Rightarrow$ (b): Since
$$X^0 = \{f \in X' : \hat{x}(f) = f(x) = 0 \quad \text{for all } x \in X\} = \{0\},$$
it follows from corollary (10.23) that $\{X\} = (X')_n^b$, and hence from statement (a) that $X = (X')_n^b$.

(b) $\Rightarrow$ (c): Note that $(X')_n^b$ is a normal subspace of $(X')^b$. It follows from the statement (b) that X is the topological dual of $(X', C', \sigma_S(X', X))$, and hence from corollary (13.10) that X is complete for $\sigma_S(X, X')$.

(c) $\Rightarrow$ (a): In view of theorems (13.9) and (13.5), X is an ℓ-ideal in $(X')_n^b$, it follows from corollary (6.5) that $X = (X', C', \sigma_S(X', X))'$, and hence from corollary (13.10) that X is a normal subspace of $(X')_n^b$.

A Riesz space (X, C) is said to be *perfect* if X_n^b is total over X and $X = X_{nn}$. Using theorem (13.9) and corollary (13.12), we give some characterizations of perfectness in terms of topological completeness. The equivalence of (a) and (e) in the following corollary was given by Nakano (1950a).

(13.13) COROLLARY. *For any Riesz space* (X, C), *if* X_n^b *is total over* X, *then the following statements are equivalent:*

(a) (X, C) *is perfect;*

(b) X *is a normal subspace of* $X_{nn};$

(c) X *is complete for* $\sigma_S(X, X_n^b);$

(d) *each positive* $\sigma(X, X_n^b)$-*bounded subset of* X *which is directed upwards has a* $\sigma(X, X_n^b)$-*limit;*

(e) $(X, C, \sigma(X, X_n^b))$ *is boundedly order-complete;*

(f) $(X, C, \sigma_S(X, X_n^b))$ *is both locally order-complete and boundedly order-complete.*

Proof. Give X the topology $\sigma_S(X, X_n^b)$. Then the equivalence of (a)–(c) is a restatement of corollary (13.12) and the equivalence of (c)–(f) follows from theorem (13.9).

(13.14) COROLLARY. *Let* (X, C) *be an order-complete Riesz space for which* X_n^b *is total over* X. *Then* X_{nn} *is the completion of* X *for* $\sigma_S(X, X_n^b)$.

Proof. By making use of corollary (13.7), X can be regarded as an ℓ-ideal in X_{nn}, hence, by corollary (6.5), X is the topological dual of the locally convex Riesz space $(X_n^b, \sigma_S(X_n^b, X))$. In view of corollary (13.11), the completion of X for $\sigma_S(X, X_n^b)$ is the normal subspace of X_{nn} generated by X, consequently, by corollary (13.7), X_{nn} is exactly the completion of X for $\sigma_S(X, X_n^b)$.

The completeness of X' for $\sigma_S(X', X)$ ensures that $\sigma_S(X', X)$ and $\mathscr{P}_b(X')$ have the same topologically bounded sets as shown by the following result.

(13.15) PROPOSITION. *Let $(X, C, \mathscr{P})$ be a locally convex Riesz space with the topological dual X', and let $\mathscr{P}_{\mathrm{b}}(X')$ be the order-bound topology on X'. Suppose that X' is a normal subspace of X^{b}. Then, for any subset B of X', the following statements are equivalent:*

(a) *B is $\mathscr{P}_{\mathrm{b}}(X')$-bounded;*
(b) *B is $\beta(X', X)$-bounded;*
(c) *B is $\sigma_{\mathrm{S}}(X', X)$-bounded.*

Proof. It is clear that $\sigma_{\mathrm{S}}(X', X)$ is coarser than $\beta(X', X)$ and that $\beta(X', X)$ is coarser than $\mathscr{P}_{\mathrm{b}}(X')$. Then the implications (a) $\Rightarrow$ (b) $\Rightarrow$ (c) are obvious. It remains to verify that (c) implies (a). Without loss of generality we can assume that B is a convex, solid, and $\sigma_{\mathrm{S}}(X', X)$-bounded subset of X'. If B is not $\mathscr{P}_{\mathrm{b}}(X')$-bounded, there exists a convex and solid $\mathscr{P}_{\mathrm{b}}(X')$-neighbourhood V of 0 in X' such that the assertion $B \subset 2^{2n} V$ is false for each natural number n. For any n, there exists f_n in X' such that $|f_n| \in B$ and $|f_n| \notin 2^{2n} V$. Let $g_n = \sum_{k=1}^{n} 2^{-k} |f_k|$. Then $\{g_n\}$ is a $\sigma_{\mathrm{S}}(X', X)$-Cauchy sequence in X' because B is convex and $\sigma_{\mathrm{S}}(X', X)$-bounded. Since X' is a normal subspace of X^{b}, by corollary (13.10) X' is complete for $\sigma_{\mathrm{S}}(X', X)$, and so there exists $g \in X'$ such that g_n converges to g with respect to $\sigma_{\mathrm{S}}(X', X)$. Since $0 \leqslant g_n\uparrow$ and since C' is $\sigma_{\mathrm{S}}(X', X)$-closed, it follows from proposition (2.1) that $g = \sup g_n$. Since V is a $\mathscr{P}_{\mathrm{b}}(X')$-neighbourhood of 0 in X', there exists a natural number k such that $[0, g] \subset 2^k V$. We conclude from $0 \leqslant 2^{-k} |f_k| \leqslant g_k \leqslant g$ that $|f_k| \in 2^{2k} V$, contradicting the fact that $|f_n| \notin 2^{2n} V$. Therefore B is $\mathscr{P}_{\mathrm{b}}(X')$-bounded, and the proof is complete.

We now present a dual result to proposition (13.15) as follows.

(13.16) COROLLARY. *Let $(X, C, \mathscr{P})$ be a locally convex Riesz space with the topological dual X', and let $\mathscr{P}_{\mathrm{b}}$ be the order-bound topology on X. Suppose that X is a normal subspace of $(X')_{\mathrm{n}}^{\mathrm{b}}$. Then, for any subset A of X, the following statements are equivalent:*

(a) *A is $\mathscr{P}_{\mathrm{b}}$-bounded;*
(b) *A is $\beta(X, X')$-bounded;*
(c) *A is $\sigma(X, X')$-bounded.*

Proof. Since X' is an ℓ-ideal in X^{b}, then $\sigma(X, X')$ and $\sigma_{\mathrm{S}}(X, X')$ are consistent with the duality $\langle X, X' \rangle$, therefore a subset A of X is $\sigma(X, X')$-bounded if and only if it is $\sigma_{\mathrm{S}}(X', X)$-bounded. On the other hand, since X is a normal subspace of $(X')_{\mathrm{n}}^{\mathrm{b}}$, X must be the topological dual of the locally convex Riesz space $(X', C', \sigma_S(X', X))$, and so the result follows from proposition (13.15).

(13.17) PROPOSITION. *Let $(X, C, \mathscr{P})$ be a locally convex Riesz space with the topological dual X'. If X' is complete for $\sigma_S(X', X)$ then it is complete for $\beta(X', X)$.*

Proof. Let $\{f_\tau : \tau \in D\}$ be a $\beta(X', X)$-Cauchy net in X'. Then it must be a $\sigma_S(X', X)$-Cauchy net because $\sigma_S(X', X)$ is clearly coarser than $\beta(X', X)$. Since X' is complete for $\sigma_S(X', X)$, the net $\{f_\tau : \tau \in D\}$ $\sigma_S(X', X)$-converges in X', say to f. (In particular, $\{f_\tau\}$ converges to f with respect to $\sigma(X', X)$.) We now show that f_τ converges to f with respect to $\beta(X', X)$. Let A be any solid, convex, and $\sigma(X, X')$-bounded subset of X. There exists $\tau_0 \in D$ such that $f_\tau - f_{\tau_0} \in \frac{1}{2}A^0$ whenever $\tau \geqslant \tau_0$, where A^0 is the polar of A taken in X'. Since A^0 is $\sigma(X', X)$-closed and since f_τ converges to f with respect to $\sigma(X', X)$, it follows that $f - f_{\tau_0} \in \frac{1}{2}A^0$, and hence that

$$f - f_\tau = (f - f_{\tau_0}) + (f_{\tau_0} - f_\tau) \in \tfrac{1}{2}A^0 + \tfrac{1}{2}A^0 = A^0$$

whenever $\tau \geqslant \tau_0$. Therefore f_τ converges to f with respect to $\beta(X', X)$; consequently X' is complete for $\beta(X', X)$. This completes the proof.

It will be seen from theorems (11.16), (18.17) and example (15.9) that the converse of the above result is not true in general.

The following corollary, which should be compared with theorem (13.9), is a dual result to proposition (13.17).

(13.18) COROLLARY. *Let $(X, C, \mathscr{P})$ be a locally convex Riesz space. If X is complete for $\sigma_S(X, X')$ then it is complete for $\beta(X, X')$ and also for $\mathscr{P}$.*

Proof. It is known from corollaries (13.12) and (6.5) that X is the topological dual of the locally convex Riesz space $(X', C', \sigma_S(X', X))$; it follows from proposition (13.17) that X is complete for $\beta(X, X')$. Finally the completeness of X for $\mathscr{P}$ follows from a well-known result.

Of course, $\sigma_S(X, X')$ is the coarsest locally solid topology on X consistent with the dual pair $\langle X, X' \rangle$ while $\beta(X, X')$ is not consistent with $\langle X, X' \rangle$, therefore the preceding corollary is of particular interest.

REFLEXIVITY FOR LOCALLY
CONVEX RIESZ SPACES

THIS chapter is concerned with a study of the interrelation between reflexivity and order. It is known from Kōmura (1964) that there are reflexive locally convex spaces which are not topologically complete. But for locally convex Riesz spaces completeness is a consequence of semi-reflexivity as shown by the following result due to Wong (1969c).

(14.1) THEOREM. *For any locally convex Riesz space* $(X, C, \mathscr{P})$, *the following statements are equivalent:*

(a) X *is semi-reflexive;*

(b) $(X, C, \sigma_S(X, X'))$ *is both locally and boundedly order-complete, and* $X'' \subset (X')^b_n;$

(c) X *is complete for* $\sigma_S(X, X')$, *and* $X'' \subset (X')^b_n;$

(d) *each positive* $\sigma(X, X')$-*bounded subset of* X *which is directed upwards has a* $\sigma(X, X')$-*limit, and* $X'' \subset (X')^b_n;$

(e) *each positive* $\mathscr{P}$-*bounded subset of* X *which is directed upwards has a* $\mathscr{P}$-*limit, and* $X'' \subset (X')^b_n;$

(f) $(X, C, \mathscr{P})$ *is boundedly order-complete,* $X' \subset X^b_n$ *and* $X'' \subset (X')^b_n;$

(g) X *is a normal subspace of* $(X')^b_n$, *and* $X'' \subset (X')^b_n.$

Furthermore, if $(X, C, \mathscr{P})$ *satisfies one of* (a)–(g) *then* X *is complete for* $\mathscr{P}$ *and also for* $\beta(X, X').$

Proof. We have only to verify the implications (a) $\Rightarrow$ (b) $\Rightarrow$ (g) $\Rightarrow$ (a). Other equivalences follow from theorem (13.9). Observe that X is always a Riesz subspace of $(X')^b_n$. It follows from the semi-reflexivity of X that $X'' = X \subset (X')^b_n$ and hence from proposition (12.10) that $(X, C, \sigma_S(X, X'))$ is locally order-complete. Let $\{u_\tau\}$ be a $\sigma_S(X, X')$-bounded subset of X which is directed upwards. Since $\sigma_S(X, X')$ is consistent with the duality $\langle X, X' \rangle$, then $\{u_\tau\}$ is $\mathscr{P}$-bounded and hence the polar $(\{u_\tau\})^0$ of the set $\{u_\tau\}$, taken in X', is a $\beta(X', X)$-neighbourhood of 0 in X'. According to the Alaoglu–Bourbaki theorem, $\{u_\tau\}$ is relatively $\sigma(X, X')$-compact, and hence $\{u_\tau\}$ has a $\sigma(X, X')$-cluster point u in X, consequently $u_\tau \uparrow u$. Therefore, $(X, C, \sigma_S(X, X'))$ is boundedly order-complete; this proves the implication (a) $\Rightarrow$ (b). The implication (b) $\Rightarrow$ (g) is a consequence of corollary (13.12). To see the implication (g) $\Rightarrow$ (a), let

$$X^0 = \{h \in X' : h(x) = 0 \quad \text{for all } x \text{ in } X\}$$

and

$$(X'')^0 = \{f \in X' : \phi(f) = 0 \quad \text{for all } \phi \text{ in } X''\}.$$

Then $X^0 = \{0\} \subseteq (X'')^0$; since X and X'' are ℓ-ideals in $(X')_n^b$ it follows from corollary (10.23) that $\{X\} \supseteq \{X''\}$. By (g), $X = \{X\}$. Consequently $X \supseteq \{X''\} \supseteq X''$ and so $X = X''$. This shows that X is semi-reflexive and proves the implication (g) $\Rightarrow$ (a).

The final assertion that X is complete for $\mathscr{P}$ and also for $\beta(X, X')$ is a consequence of corollary (13.18), therefore the proof is complete.

Remark. Since X'' is the topological dual of the locally convex Riesz space $(X', C', \beta(X', X))$, then the condition that $X'' \subset (X')_n^b$ in theorem (14.1) can be replaced by any one of the equivalent properties of theorem (13.1).

Since semi-reflexivity and reflexivity are equivalent for normed vector spaces, we obtain an immediate consequence of the preceding result as follows.

(14.2) COROLLARY (Ogasawara). *Let $(X, C, \|\,.\,\|)$ be a normed Riesz space. Then the following statements are equivalent and each is equivalent to each of (b)–(g) of theorem (14.1):*

(i) *$(X, C, \|\,.\,\|)$ is reflexive;*

(ii) *$(X, C, \|\,.\,\|)$ is boundedly order-complete, $X' \subset X^b$ and $X'' \subset (X')_n^b$.*

(14.3) COROLLARY. *Let $\mathscr{T}$ be a locally convex topology on a Riesz space (X, C) such that the topological dual X' of $(X, \mathscr{T})$ is an ℓ-ideal in X^b. Then the following statements are equivalent and each is equivalent to each of (b)–(g) of theorem (14.1):*

(a) *X is semi-reflexive;*

(b) *X is complete for $\sigma_S(X, X')$, and $X'' \subset (X')_n^b$.*

Proof. Since X' is an ℓ-ideal in X^b, it follows that $\sigma(X, X')$ is a locally o-convex topology on X, and hence that $\sigma_S(X, X')$ and $\mathscr{T}$ are consistent with the duality $\langle X, X' \rangle$. On the other hand, since semi-reflexivity and boundedly order-completeness depend only on the duality, the result now follows from theorem (14.1).

(14.4) COROLLARY. *Let (X, C) be a Riesz space, X_c^b total over (X, C), and let $\mathcal{T}$ be a locally convex topology (not necessarily locally solid) on X which is consistent with the duality $\langle X, X_c^b \rangle$. Then the following statements are equivalent and each is equivalent to* (b)–(g) *of theorem* (14.1):
  (a) X *is semi-reflexive;*
  (b) $X_c^b = X_n^b$, X *is perfect, and* $(X_c^b, \beta(X_c^b, X))' \subset X_{nn}$.

Proof. (a) $\Rightarrow$ (b): Let σ_S denote the topology $\sigma_S(X, X_c^b)$. Then $(X, \sigma_S)' = X_c^b$. By (a), (X, σ_S) is semi-reflexive. By the implication (a) $\Rightarrow$ (f) of theorem (14.1), (X, C, σ_S) is boundedly order-complete, $X_c^b \subset X_n^b$ and $(X^b, \beta(X_c^b, X_c^b))' \subset (X_c^b)_n^b$. Consequently, $X_c^b = X_n^b$ and $(X_c^b, \beta(X_c^b, X))' \subset (X_n^b)_n^b = X_{nn}$. Further, by (c) of theorem (14.1), X is $\sigma_S(X, X_n^b)$-complete and it follows from corollary (13.13) that X is perfect.

(b) $\Rightarrow$ (a): By the first two properties stated in (b), it follows from corollary (13.13) that (X, σ_S) is complete. The third property stated in (b) implies that $(X_c^b, \beta(X_c^b, X))' \subset (X_c^b)_n$. By the implication (c) $\Rightarrow$ (a) of theorem (14.1), we conclude that X is semi-reflexive.

(14.5) COROLLARY. *Let (X, C) be a Riesz space, and let $\mathcal{T}$ be a locally convex topology on X such that the topological dual X' of $(X, \mathcal{T})$ is an ℓ-ideal in X^b. If $(X, \mathcal{T})$ is reflexive then $\mathcal{T}$ is a locally solid topology, consequently X is complete for $\sigma_S(X, X')$ and also for $\mathcal{T}$.*

Proof. It is known that $\sigma(X, X')$ is a locally o-convex topology on X. Since X' is an ℓ-ideal in X^b, it follows from corollary (6.5) that $(X, C, \sigma_S(X, X'))' = X'$, and hence from theorem (11.16) that the strong dual $(X', C', \beta(X', X))$ of $(X, \mathcal{T})$ is a locally convex Riesz space. Since $(X, C, \mathcal{T})$ is reflexive, $(X, C, \mathcal{T})$ is the strong dual of $(X', C', \beta(X', X))$ so, by theorem (11.16) again, $(X, C, \mathcal{T})$ is a locally convex Riesz space, consequently $\mathcal{T}$ is a locally solid topology. Finally, in view of theorem (14.1), X is complete for $\sigma_S(X, X')$ and also for $\mathcal{T}$.

Theorem (14.1) leads to the following characterizations of reflexivity for locally convex Riesz spaces.

(14.6) THEOREM. *For any locally convex Riesz space $(X, C, \mathcal{P})$ with the topological dual X', the reflexivity of $(X, C, \mathcal{P})$ is equivalent to the following three conditions:*

(a) $\mathscr{P}$ is the Mackey topology $\tau(X, X')$;

(b) if $u_\tau \downarrow 0$ in X then u_τ converges to 0 with respect to $\beta(X, X')$, and if $f_\tau \downarrow 0$ in X' then f_τ converges to 0 with respect to $\beta(X', X)$;

(c) $(X, C, \sigma(X, X'))$ and $(X', C', \sigma(X', X))$ are boundedly order-complete.

(*Remark.* The statement that if $u_\tau \downarrow 0$ in X then u_τ converges to 0 with respect to $\beta(X, X')$ in theorem (14.6)(b) can be replaced by one of (a)–(g) of theorem (13.1); similarly for the statement that if $f_\tau \downarrow 0$ in X' then f_τ converges to 0 with respect to $\beta(X', X)$ in theorem (14.6)(b). Also the statement that $(X, C, \sigma(X, X'))$ is boundedly order-complete in theorem (14.6)(c) is equivalent to each of (a)–(e) of theorem (13.9), and there is a similar equivalence for the statement that $(X', C' \sigma(X', X))$ is boundedly order-complete in theorem (14.6)(c).)

Proof. It is well known that $(X, C, \mathscr{P})$ is reflexive if and only if $\mathscr{P}$ is $\tau(X, X')$ and both $(X, C, \sigma(X, X'))$ and $(X', C', \beta(X', X))$ are semi-reflexive. Then the necessity follows from theorems (14.1), (13.5), and (13.1) and $\mathscr{P} = \beta(X, X')$. Conversely, if $u_\tau \downarrow 0$ in X, then u_τ must converge to 0 with respect to $\mathscr{P}$ since $\mathscr{P}$ is, in general, coarser than $\beta(X, X')$ and hence $(X, C, \sigma(X, X'))$ is semi-reflexive in view of theorems (13.9) and (13.1); on the other hand, because of corollary (13.10) and the fact that $X'' = X \subseteq (X')^{\mathrm{b}}_{\mathrm{n}}$, $(X', C', \beta(X', X))$ must be semi-reflexive. Therefore $(X, C, \mathscr{P})$ is reflexive, and the proof is complete.

Let $\langle E_1, E_2 \rangle$ be a dual pair. Recall that E_1 is said to be *semi-reflexive with respect to* E_2 if $E_1 = (E_2, \beta(E_2, E_1))'$ and that the dual pair $\langle E_1, E_2 \rangle$ is said to be *reflexive* if E_1 is semi-reflexive with respect to E_2 and E_2 is semi-reflexive with respect to E_1. For further information about the reflexivity of a dual pair, we refer the reader to Köthe (1969).

(14.7) PROPOSITION. *For any Riesz space (X, C), if $X^{\mathrm{b}}_{\mathrm{n}}$ is total over X then X is semi-reflexive with respect to $X^{\mathrm{b}}_{\mathrm{n}}$ if and only if the following two conditions hold:*

(a) X *is perfect;*

(b) *if* $f_\tau \downarrow 0$ *in* $X^{\mathrm{b}}_{\mathrm{n}}$ *then* f_τ *converges to 0 with respect to* $\beta(X^{\mathrm{b}}_{\mathrm{n}}, X)$.

Proof. Since $X^{\mathrm{b}}_{\mathrm{n}}$ is a normal subspace of X^{b}, it follows from corollary (6.5) that $X^{\mathrm{b}}_{\mathrm{n}} = (X, C, \sigma_{\mathrm{S}}(X, X^{\mathrm{b}}_{\mathrm{n}}))'$. Therefore the semi-reflexivity of X with respect to $X^{\mathrm{b}}_{\mathrm{n}}$ is equivalent to the semi-reflexivity of $(X, C, \sigma_S(X, X_n^b))$, and this is the case if and only if X is complete for $\sigma_S(X, X_n^b)$ and $(X_n^b, \beta(X_n^b, X))' \subset X_{nn}$; consequently the result now follows from theorem (13.1) and corollary (13.13).

Remark. The condition (a) in proposition (14.7) can be replaced by one of (a)–(f) of corollary (13.13), and the condition (b) in proposition (14.7) is equivalent to each of (a)–(g) of theorem (13.1).

(14.8) COROLLARY. *For any Riesz space (X, C), if X_n^b is total over X, then the dual pair $\langle X, X_n^b \rangle$ is reflexive if and only if the following two conditions hold:*

(a) *X is perfect;*

(b) *if $u_\tau \downarrow 0$ in X then u_τ converges to 0 with respect to $\beta(X, X_n^b)$ and if $f_\tau \downarrow 0$ in X_n^b then f_τ converges to 0 with respect to $\beta(X_n^b, X)$.*

Proof. The necessity is obvious. To prove the sufficiency, we first note that the perfectness of X_n^b is a direct consequence of the perfectness of X and so, by proposition (14.7), X_n^b is also semi-reflexive with respect to X. Consequently the dual pair $\langle X, X_n^b \rangle$ is reflexive.

BORNOLOGICAL AND
INFRABARRELLED RIESZ SPACES

THE remaining four chapters of this book are devoted to studying some important classes of locally convex Riesz spaces and their relationship. Let $(E, \mathscr{P})$ be a locally convex space. Recall that $(E, \mathscr{P})$ is bornological if each convex circled subset of E which absorbs all $\mathscr{P}$-bounded subsets of E is a $\mathscr{P}$-neighbourhood of 0, $(E, \mathscr{P})$ is barrelled if each barrel in E is a $\mathscr{P}$-neighbourhood of 0, and that $(E, \mathscr{P})$ is infrabarrelled if each barrel in E which absorbs all $\mathscr{P}$-bounded subsets of E is a $\mathscr{P}$-neighbourhood of 0. Clearly bornological spaces and barrelled spaces are infrabarrelled. A locally convex Riesz space $(X, C, \mathscr{P})$ is called a *bornological Riesz space* if $(X, \mathscr{P})$ is bornological; it is called an *infrabarrelled Riesz space* if $(X, \mathscr{P})$ is infrabarrelled. It is known from proposition (11.2)(c) that if $\mathscr{P}$ is a locally solid topology on (X, C) then the solid hull of each $\mathscr{P}$-bounded subset of X is $\mathscr{P}$-bounded. Therefore the question arises naturally whether the converse of proposition (11.2)(c) is true, namely: If $\mathscr{P}$ is a locally convex topology on (X, C) such that the solid hull of each $\mathscr{P}$-bounded set in E is $\mathscr{P}$-bounded, is $\mathscr{P}$ a locally solid topology? Example (3.15) shows that the answer to the above question is, in general, negative. However, we have the following theorem.

(15.1) THEOREM. *Let $\mathscr{P}$ be a locally convex topology on (X, C) such that $(X, \mathscr{P})$ is bornological. If the solid hull of each $\mathscr{P}$-bounded subset of X is $\mathscr{P}$-bounded, then $\mathscr{P}$ is a locally solid topology.*

Proof. We first note that each order-bounded subset of X is $\mathscr{P}$-bounded because C is generating and the solid hull of each $u \in C$ is $[-u, u]$. Let $\mathscr{U}$ be a neighbourhood-base at 0 for $\mathscr{P}$ consisting of convex and circled sets in X, and suppose that

$$\tilde{\mathscr{U}} = \{\mathrm{sk}(V) : V \in \mathscr{U}\},$$

where $\mathrm{sk}(V)$ is the solid kernel of V. In view of proposition (10.5)(d), each $\mathrm{sk}(V)$ absorbs all order-bounded subsets of X, and is certainly absorbing. On the other hand, it is easily seen that $\mathrm{sk}(\lambda V) = \lambda \, \mathrm{sk}(V)$

for all $\lambda \neq 0$, $$\text{sk}(V) \cap \text{sk}(W) = \text{sk}(V \cap W),$$

and that $\text{sk}(W) \subseteq \text{sk}(V)$ whenever $W \subseteq V$. There exists a unique locally solid topology, $\mathscr{T}$ say, on X such that $\tilde{\mathscr{U}}$ is a neighbourhood-base at 0 for $\mathscr{T}$. Clearly $\mathscr{T}$ is the greatest lower bound of all locally solid topologies on X which are finer than $\mathscr{P}$.

We now claim that any $\mathscr{P}$-bounded subset of X is $\mathscr{T}$-bounded. In fact, let A be a $\mathscr{P}$-bounded subset of X. Then A is contained in a solid $\mathscr{P}$-bounded subset B of X. For this B, there exists V in $\mathscr{U}$ such that $B \subseteq nV$ for some natural number n. Since B is solid and $\text{sk}(V)$ is the largest solid subset of V under the inclusion, it follows that $A \subseteq B \subseteq n \, \text{sk}(V)$ for this n. This shows that A is $\mathscr{T}$-bounded.

Finally, since $(X, \mathscr{P})$ is bornological or, equivalently, no strictly finer locally convex topology on X has the same topologically bounded sets, we conclude that $\mathscr{P}$ and $\mathscr{T}$ coincide, and hence that $\mathscr{P}$ is locally solid. This completes the proof.

(15.2) COROLLARY. *Let $\mathscr{P}$ be a locally convex topology on (X, C), and let X^{tb} be the topologically bounded dual of $(X, \mathscr{P})$. If the solid hull of each $\mathscr{P}$-bounded subset of X is $\mathscr{P}$-bounded then the Mackey topology $\tau(X, X^{\text{tb}})$ is a locally solid topology on X. In particular, the bornological space associated with a locally convex Riesz space is always a locally convex Riesz space.*

Proof. Since the locally convex topologies $\mathscr{P}$ and $\tau(X, X^{\text{tb}})$ have the same topologically bounded sets in X, it follows that the solid hull of each $\tau(X, X^{\text{tb}})$-bounded subset of X is $\tau(X, X^{\text{tb}})$-bounded, and hence the result follows from theorem (15.1).

A semi-norm p on an ordered topological vector space is said to be *topologically bounded* if it sends every topologically bounded set to a bounded set; it is said to be *order-bounded* if it sends every order-bounded set to a bounded set. In terms of the order structure, we are able to give some characterizations of bornological Riesz spaces as follows.

(15.3) PROPOSITION. *For any locally convex Riesz space $(X, C, \mathscr{P})$ the following statements are equivalent:*

(a) *$(X, C, \mathscr{P})$ is bornological;*

(b) *each decomposable circled convex set in X which absorbs all $\mathscr{P}$-bounded subsets of X is a $\mathscr{P}$-neighbourhood of 0;*

(c) *each circled o-convex set in X which absorbs all $\mathscr{P}$-bounded subsets of X is a $\mathscr{P}$-neighbourhood of 0;*

(d) *each convex solid set in X which absorbs all $\mathscr{P}$-bounded subsets of X is a $\mathscr{P}$-neighbourhood of 0;*

(e) *each topologically bounded, Riesz semi-norm p on X is $\mathscr{P}$-continuous.*

(*Remark.* Before giving the proof of this proposition, we note that, since C is a strict $\mathscr{B}$-cone in $(X, \mathscr{P})$, a subset B of X absorbs all $\mathscr{P}$-bounded subsets of X if and only if it absorbs all positive $\mathscr{P}$-bounded sets in X, therefore, in the statements (b), (c), and (d) of the proposition, '$\mathscr{P}$-bounded subsets of X' may be replaced by 'positive $\mathscr{P}$-bounded subsets of X'.)

Proof. The implication (a) $\Rightarrow$ (b) is trivial. It is noted that for any convex circled set V in X, $\mathrm{co}(-(V \cap C) \cup (V \cap C))$ is a decomposable subset of X for which

$$\mathrm{co}(-(V \cap C) \cup (V \cap C)) \subseteq V,$$

and
$$C \cap \{\mathrm{co}(-(V \cap C) \cup (V \cap C))\} = V \cap C.$$

It then follows that (b) implies (c). Suppose that V is a convex and solid subset of X and that W is the order-convex hull of B, i.e.

$$W = (B+C) \cap (B-C),$$

where $B = \mathrm{co}(-(V \cap C) \cup (V \cap C))$. Then $W \cap C = V \cap C = B \cap C$, $W \subseteq 2V$, and W is o-convex and circled; consequently V absorbs all $\mathscr{P}$-bounded subsets of X if and only if W absorbs all $\mathscr{P}$-bounded subsets of X since C is a strict $\mathscr{B}$-cone in $(X, \mathscr{P})$; consequently (c) implies (d). Since a subset V of X absorbs each $\mathscr{P}$-bounded set in X if and only if the solid kernel $\mathrm{sk}(V)$ of V absorbs all $\mathscr{P}$-bounded sets in X, it follows that (d) implies (a). Observe that the implication (d) $\Rightarrow$ (e) follows from the fact that $W = \{x \in X : p(x) \leqslant 1\}$ is a convex solid subset of X which absorbs all $\mathscr{P}$-bounded subsets of X. Finally, since the gauge of a convex, solid, and absorbing subset of X must be a Riesz semi-norm, then (e) implies (d). Therefore the proof is complete.

It is well known that subspaces of a bornological, locally convex space are, in general, not bornological with respect to the relative topology. But the following result shows that any ℓ-ideal in a bornological Riesz space must be bornological with respect to the relative topology.

(15.4) Corollary. *Any ℓ-ideal in a bornological Riesz space is bornological with respect to the relative topology.*

Proof. Let J be an ℓ-ideal in a bornological Riesz space $(X, C, \mathscr{P})$, and let V be a convex solid subset of J which absorbs all $\mathscr{P}$-bounded subsets of J. In view of the preceding proposition, we have only to show that V is a neighbourhood of 0 in the subspace J.

Let
$$U = \{x \in X : y \in V \quad \text{whenever } 0 \leqslant y \leqslant |x| \text{ and } y \in J\}.$$

Then U is a convex solid set in X such that $U \cap J = V$. Further, U absorbs all $\mathscr{P}$-bounded subsets B of X. To verify this, we may assume without loss of generality that B is solid. Now, if $B \not\subseteq nU$ for each positive integer n, then there exists $b_n \in B$ such that $b_n \notin nU$. Hence there is $y_n \in J$ with $0 \leqslant y_n \leqslant |b_n|/n|$ such that $y_n \notin V$. Since B is solid, the set $\{ny_n\}_{n=1}^{\infty}$ is contained in $B \cap J$, but fails to be absorbed by V, contrary to our assumption on V. Therefore U must absorb all $\mathscr{P}$-bounded sets in X. Since X is bornological, it follows that U is a $\mathscr{P}$-neighbourhood of 0 in X. Since $V = U \cap J$, V is a neighbourhood of 0 in J with respect to the relative topology.

Similar to the case of bornological spaces, we shall show that each bornological Riesz space is the inductive limit of a family of normed Riesz spaces with respect to ℓ-homomorphisms.

(15.5) Proposition. *Every bornological Riesz space $(X, C, \mathscr{P})$ is the inductive limit of a family of normed Riesz spaces (and of B-lattices if X is quasi-complete for $\mathscr{P}$) with respect to ℓ-homomorphisms; the cardinality of this family can be so chosen as the cardinality of any fundamental system of $\mathscr{P}$-bounded sets in X.*

Proof. Let $\mathscr{B}$ denote the family of all $\mathscr{P}$-bounded, convex, solid, and $\mathscr{P}$-closed sets in X. According to propositions (11.2) and (11.3), $\mathscr{B}$ is a fundamental system of the family consisting of all $\mathscr{P}$-bounded subsets of X. For each $B \in \mathscr{B}$, suppose that $X_B = \bigcup_n nB$ and that p_B is the gauge of B on X_B. Since B is solid and convex, X_B is an ℓ-ideal in X, and p_B is a Riesz semi-norm on X_B. Observe that the relative topology on X_B induced by $\mathscr{P}$ is coarser than the semi-norm topology p_B; it follows that (X_B, C_B, p_B) is a normed Riesz space, where $C_B = C \cap X_B$. Obviously each injection j_B of X_B into X is an ℓ-homomorphism, and $X = \cup \{X_B : B \in \mathscr{B}\}$. By a well-known result (cf.

13

Schaefer (1966)), $\mathscr{P}$ is the inductive topology with respect to the families $\{(X_B, p_B) : B \in \mathscr{B}\}$ and $\{j_B : B \in \mathscr{B}\}$, therefore $(X, C, \mathscr{P})$ is the inductive limit of a family $\{(X_B, C_B, p_B) : B \in \mathscr{B}\}$ of normed Riesz spaces with respect to ℓ-homomorphisms $\{j_B : B \in \mathscr{B}\}$. This completes the proof.

A subset V of a locally convex Riesz space $(X, C, \mathscr{P})$ is called a *solid barrel* if it is $\mathscr{P}$-closed, convex, solid, and absorbing. It is known from proposition (11.3) that a barrel V absorbs all $\mathscr{P}$-bounded subsets of X if and only if the solid barrel $\mathrm{sk}(V)$ absorbs all $\mathscr{P}$-bounded sets in X. This deduces the following characterization of infrabarrelled Riesz spaces.

(15.6) PROPOSITION. *For any locally convex Riesz space* $(X, C, \mathscr{P})$, *the following statements are equivalent:*

(a) $(X, C, \mathscr{P})$ *is infrabarrelled;*

(b) *each solid barrel in X which absorbs all positive $\mathscr{P}$-bounded subsets of X is a $\mathscr{P}$-neighbourhood of 0;*

(c) *each solid barrel in X which absorbs all $\mathscr{P}$-bounded subsets of X is a $\mathscr{P}$-neighbourhood of 0;*

(d) *each topologically bounded lower semi-continuous Riesz semi-norm on X is $\mathscr{P}$-continuous;*

(e) *each positive $\beta(X', X)$-bounded subset of X' is $\mathscr{P}$-equicontinuous.*

The proof of this result is similar to that given in proposition (15.3) and will be left to the reader.

(15.7) COROLLARY. *Any ℓ-ideal in an infrabarrelled Riesz space is infrabarrelled with respect to the relative topology.*

Proof. Let J be an ℓ-ideal in an infrabarrelled Riesz space $(X, C, \mathscr{P})$ and let V be a solid barrel in the subspace J which absorbs all $\mathscr{P}$-bounded subsets of J. Let

$$U = \{x \in X : y \in V \quad \text{whenever } 0 \leqslant y \leqslant |x| \text{ and } y \in J\}.$$

As in the proof of corollary (15.4), U is a convex solid set in X such that $V = U \cap J$ and U absorbs all $\mathscr{P}$-bounded subsets of X. Further, U is $\mathscr{P}$-closed. In fact, let $x \in \bar{U}$ and suppose that $\{x_\tau\}$ is a net in U which is $\mathscr{P}$-convergent to x. Let $y \in J$ be such that $0 \leqslant y \leqslant |x|$. Then $0 \leqslant y \wedge |x_\tau| \leqslant |x_\tau| \in U$ and each $y \wedge |x_\tau| \in J$ since J is a ℓ-ideal. By the definition of U, it follows that $y \wedge |x_\tau| \in V$. By the continuity of the lattice operations, we conclude that $y \wedge |x_r|$ converges to $y \wedge |x| = y$; hence $y \in \bar{V} = V$. This shows that $x \in U$ and hence that U is $\mathscr{P}$-closed. Therefore U is a solid barrel in X which absorbs all $\mathscr{P}$-bounded sets in X. Hence U is a $\mathscr{P}$-neighbourhood of 0 in X. Consequently, $V = U \cap J$ is a neighbourhood of 0 in the subspace J. This implies that J is infrabarrelled.

It should be noted that the quotient Riesz space and the locally convex direct sum of infrabarrelled Riesz spaces are infrabarrelled; the quotient Riesz space and the locally convex direct sum of bornological Riesz spaces are bornological.

Combining propositions (15.3) and (15.6) we have the following result.

(15.8) PROPOSITION. *Let $(X, C, \mathscr{P})$ be an infrabarrelled Riesz space. Then the following statements are equivalent:*

(a) *$(X, C, \mathscr{P})$ is bornological;*

(b) *each topologically bounded monotonic semi-norm on X is lower semi-continuous;*

(c) *each topologically bounded Riesz semi-norm on X is lower semi-continuous.*

It is known from proposition (7.1) that if $\mathscr{P}_b$ is the order-bound topology on (X, C) then $(X, C, \mathscr{P}_b)$ is bornological. The question naturally arises whether the topology on a bornological Riesz space is necessarily the order-bound topology. The following example gives a negative answer.

(15.9) EXAMPLE. Let X be the Banach space of all continuous real-valued functions on $[0, 1]$ with the supremum norm defined by $\|x\| = \max\{|x(t)| : t \in [0, 1]\}$ and let

$$C = \{x \in X : x(t) \geqslant 0 \quad \text{for all } t \in [0, 1]\}.$$

Then $(X, C, \|.\|)$ is a Banach lattice. Suppose that J is the vector subspace consisting of all elements x in X which vanish in a neighbourhood (depending on x) of $t = 0$, and that $C_r = C \cap J$. Then J is an ℓ-ideal in X, and $(J, C_r, \|.\|)$ is a normed Riesz space and *a fortiori* a bornological Riesz space. Let

$$V = \{x \in J : n\, |x(n^{-1})| \leqslant 1 \text{ for all natural numbers } n \geqslant 1\}.$$

Then V has the following properties:

(a) V is not a $\|.\|$-neighbourhood of 0 in J;

(b) V is a solid barrel in $(J, C_r, \|.\|)$.

In fact, in order to verify the assertion (a), it is sufficient to show that for any natural number $R \geqslant 1$, there exists x_R in J with $\|x_R\| \leqslant 1/R$ such that $x_R \notin V$; it then follows that 0 is not an interior point of V; consequently V is not a $\|.\|$-neighbourhood of 0. Consider two closed disjoint subsets $[0, 1/4R]$ and $[1/(R+1), 1]$ of $[0, 1]$; by Urysohn's lemma, there exists a continuous real-valued function, x_R say, on $[0, 1]$ with range in $[0, 1/R]$ such that

$$x_R(t) = 0 \quad (t \in [0, 1/4R]) \quad \text{and} \quad x_R(t) = 1/R \quad (t \in [1/(R+1), 1]).$$

Clearly $x_R \in J$ and $\|x_R\| \leqslant 1/R$. On the other hand, since

$$x_R\left(\frac{1}{R+1}\right) = \frac{1}{R} > \frac{1}{R+1},$$

it follows that $x_R \notin V$. This proves our assertion (a). It is easily seen that V is convex and $\|.\|$-closed. Let x be in J, and let α_x with $0 < \alpha_x < 1$ be such that $x(t) = 0$ for all $t \in [0, \alpha_x]$. If we choose $\lambda = \alpha_x \|x\|^{-1}$, then $\lambda x \in V$, and so V is absorbing. It remains to show that V is solid. Let x belong to V, and let y in J be such that $|y| \leqslant |x|$. It follows from

$$n |y(n^{-1})| \leqslant |x(n^{-1})| \leqslant 1$$

that $y \in V$, and hence that V is solid. This proves the second assertion. Finally since a solid set is absorbing if and only if it absorbs all order-bounded sets, this fact implies that V is a $\mathscr{P}_b$-neighbourhood of 0 in J, and hence from (a) that the norm topology on J is not the order-bound topology $\mathscr{P}_b$.

We are now in a position to establish some necessary and sufficient conditions for the topology on a bornological Riesz space to be the order-bound topology.

(15.10) PROPOSITION. *For any bornological Riesz space* $(X, C, \mathscr{P})$, *the following statements are equivalent:*
   (a) $\mathscr{P}$ *coincides with the order-bound topology* $\mathscr{P}_b$:
   (b) *each order-bounded semi-norm on* X *is topologically bounded;*
   (c) *each monotone semi-norm on* X *is topologically bounded;*
   (d) *each Riesz semi-norm on* X *is topologically bounded;*
   (e) *each positive* $\sigma(X^b, X)$-*bounded subset of* X^b *is* $\mathscr{P}$-*equicontinuous.*

Proof. The implications (b) $\Rightarrow$ (c) $\Rightarrow$ (d) are clear. If p is an order-bounded semi-norm on X, then the semi-norm $\tilde{p}$, defined by

$$\tilde{p}(x) = \sup\{p(y) : 0 \leqslant y \leqslant |x|\}, \qquad (x \in X),$$

is a Riesz semi-norm on X for which $p(x) \leqslant 2\tilde{p}(x)$ for all $x \in X$. There-fore, if the statement (d) holds then each order-bounded semi-norm on X must be topologically bounded; consequently the statements (b), (c), and (d) are equivalent. On the other hand, if the statement (e) holds, then $X^b = X'$ and so $\mathscr{P} = \tau(X, X^b)$ because $(X, C, \mathscr{P})$ is bornological; thus (e) implies (a). It remains for us to verify the implications (a) $\Rightarrow$ (b) $\Rightarrow$ (e). For any order-bounded semi-norm q on X, the set $V = \{x \in X : q(x) \leqslant 1\}$ is convex, circled, and also absorbs all order-bounded subsets of X. Now if $\mathscr{P} = \mathscr{P}_b$, then V is a $\mathscr{P}$-neighbourhood of 0, hence q is $\mathscr{P}$-continuous, and *a fortiori* topologically bounded, therefore (a) implies (b).

(b) $\Rightarrow$ (e): Let B be a positive $\sigma(X^b, X)$-bounded subset of X^b and

$$V = \{x \in X : f(|x|) < 1 \quad \text{for all } f \in B\}.$$

Then V is a solid convex and absorbing set in X, and hence absorbs all order-bounded sets in X. By (b), it follows easily that V absorbs all topologically bounded subsets of X and hence must be a $\mathscr{P}$-neighbourhood of 0 since $\mathscr{P}$ is bornological.

Corollary (15.2) leads naturally to the following question: Under what conditions is the order-bound topology on a locally convex Riesz space $(X, C, \mathscr{P})$ the topology on the bornological space associated with $(X, C, \mathscr{P})$? By using the preceding result, we are able to give an answer as follows.

(15.11) COROLLARY. *Let* $(X, C, \mathscr{P})$ *be a locally convex Riesz space with the topologically bounded dual* X^{tb}. *Then the following statements are equivalent:*

(a) $X^{tb} = X^b$;

(b) $\tau(X, X^{tb})$ *is the order-bound topology* $\mathscr{P}_b$ *on* X;

(c) *each order-bounded semi-norm on* X *is topologically bounded;*

(d) *each monotone semi-norm on* X *is topologically bounded;*

(e) *each Riesz semi-norm on* X *is topologically bounded;*

(f) *each positive* $\sigma(X^b, X)$-*bounded subset of* X^b *is* $\tau(X, X^{tb})$-*equicontinuous.*

Proof. It should be noted that $(X, C, \tau(X, X^{tb}))$ is a bornological Riesz space by corollary (15.2), and that the locally solid topologies $\mathscr{P}$ and $\tau(X, X^{tb})$ have the same topologically bounded sets. Therefore the result follows immediately from the preceding proposition.

(15.12) COROLLARY. *Let $(X, C, \mathscr{P})$ be a locally convex Riesz space with the topologically bounded dual X^{tb}. If each increasing $\mathscr{P}$-Cauchy sequence in X has an upper bound (in particular, if X is monotonically sequentially $\mathscr{P}$-complete), then $X^{tb} = X^b$. If, in addition, $(X, \mathscr{P})$ is bornological then $\mathscr{P}$ is the order-bound topology.*

Proof. If p is a Riesz semi-norm on X which is not topologically bounded, then there exists a positive, $\mathscr{P}$-bounded, convex subset B of X such that $0 \in B$ and $p(w_n) > 2^{2n}$ for all $n \geqslant 1$, where $w_n \in B$. It is clear that $\sum_{k=1}^{n} 2^{-k} w_k$ is an increasing $\mathscr{P}$-Cauchy sequence in X; and so there exists $w \in X$ such that $\sum_{k=1}^{n} 2^{-k} w_k \leqslant w$. We now conclude from

$$p(w) \geqslant 2^{-n} p(w_n) > 2^n \quad \text{for all } n \geqslant 1$$

that p is not bounded on the order-interval $[0, w]$, which gives a contradiction. Therefore, in view of corollary (15.11), $X^{tb} = X^b$, and the proof is complete.

16

THE STRUCTURE OF ORDER-
INFRABARRELLED RIESZ SPACES
AND ITS SIMPLEST PROPERTIES

I T is known that a locally convex space equipped with the finest locally convex topology is barrelled, and that the order-bound topology on a Riesz space is the finest locally solid topology. This suggests the following question: If $\mathscr{P}_b$ is the order-bound topology on (X, C), is $(X, C, \mathscr{P}_b)$ barrelled? Unfortunately the following example shows that a locally convex Riesz space, equipped with the order-bound topology, may not be barrelled.

(16.1) EXAMPLE. Let ℓ^∞ be the Banach lattice of all bounded real sequences, with the pointwise ordering and the supremum-norm $\|.\|$. For $n = 1, 2,...$ let e_n be the sequence having 1 in the nth coordinate and 0 elsewhere. The subspace E_0 of ℓ^∞ generated by the e_n consists of all finite sequences. Let e be the sequence having 1 in every coordinate, and let
$$E = \{x+\lambda e : x \in E_0, \lambda \in \mathbf{R}\},$$
with the supremum-norm $\|.\|$ and the ordering inherited from ℓ^∞. Then E is an order-unit normed space with order unit e, and so the norm topology $\|.\|$ is the order-bound topology. Also it is easily seen that E is a Riesz space. We show that E is not barrelled. Let
$$V_0 = \{(a_1, a_2,...) \in E_0 : |a_n| \leqslant 1/n \quad \text{for each } n\},$$
and let
$$V = \{x+\lambda e : x \in V_0, |\lambda| \leqslant 1\}.$$

Then V_0 is a relatively closed, convex, circled, and absorbing subset of E_0, and contains $(1/n)e_n$ for all n. Since $[-1, 1]$ is a compact subset of $\mathbf{R}$, it follows that V is a barrel in E. We show that V is not a $\|.\|$-neighbourhood of 0 in E. Let $\varepsilon > 0$. Choose a positive integer m such that $1/m < \varepsilon$ and put
$$y = \varepsilon e_m - \varepsilon e_{m+1}.$$

Then $\|y\| = \varepsilon$. We claim that $y \notin V$; suppose not, there exists $x \in V_0$ and $\lambda \in [-1, 1]$ such that $y = x+\lambda e$. Since $x = \varepsilon e_m - \varepsilon e_{m+1} - \lambda e \in V_0$

by considering the mth and $(m+1)$th coordinates of x, we have

$$|\varepsilon - \lambda| \leqslant \frac{1}{m} \qquad |-\varepsilon - \lambda| \leqslant \frac{1}{m+1}.$$

It then follows that

$$\varepsilon \leqslant \frac{1}{2}\left(\frac{1}{m} + \frac{1}{m+1}\right) \leqslant \frac{1}{m},$$

contrary to the choice of m. This shows that $y \notin V$. Since $\|y\| = \varepsilon$ and since ε is arbitrary, it follows that V is not a $\|.\|$-neighbourhood of 0 in E. Therefore, E is not barrelled.

(16.2) DEFINITION.   A locally convex Riesz space $(X, C, \mathscr{P})$ is called an *order-infrabarrelled Riesz space* if each barrel in $(X, \mathscr{P})$ which absorbs all order-bounded subsets of X is a $\mathscr{P}$-neighbourhood of 0.

It should be noted that if $(X, C, \mathscr{P})$ is a locally convex Riesz space with the topological dual X' and if $\sigma_S(X', X)$ is the locally solid topology on X' associated with $\sigma(X', X)$, then $\sigma_S(X', X)$ is coarser than the strong topology $\beta(X', X)$ because each order-bounded subset of X is $\mathscr{P}$-bounded; consequently each $\beta(X', X)$-bounded subset of X' is $\sigma_S(X', X)$-bounded.

(16.3) THEOREM.   *Let $(X, C, \mathscr{P})$ be a locally convex Riesz space with the topological dual X'. Then the following statements are equivalent:*
   (a) *$(X, C, \mathscr{P})$ is order-infrabarrelled;*
   (b) *each solid barrel in $(X, C, \mathscr{P})$ is a $\mathscr{P}$-neighbourhood of 0;*
   (c) *every order-bounded, lower semi-continuous semi-norm on X is $\mathscr{P}$-continuous;*
   (d) *every lower semi-continuous Riesz semi-norm on X is $\mathscr{P}$-continuous;*
   (e) *each barrel in $(X, \mathscr{P})$ which absorbs all relative uniform null-sequences in X is a $\mathscr{P}$-neighbourhood of 0;*
   (f) *each $\sigma_S(X', X)$-bounded subset of X' is $\mathscr{P}$-equicontinuous;*
   (g) *each positive $\sigma(X', X)$-bounded subset of X' is $\mathscr{P}$-equicontinuous.*

   Proof. A semi-norm p on X is order-bounded and lower semi-continuous if and only if the set $V = \{x \in X : p(x) < 1\}$ is a barrel in $(X, C, \mathscr{P})$ which absorbs all order-bounded subsets of X; this means that (a) and (c) are equivalent. A semi-norm q on X is a lower semi-continuous Riesz semi-norm if and only if the set $U = \{x \in X : q(x) < 1\}$ is a solid barrel in $(X, C, \mathscr{P})$; it then follows that (b) and (d) are equivalent. On the other hand, a subset W of X is a barrel in $(X, C, \mathscr{P})$ which absorbs all order-bounded sets in X if and only if the polar W^0 of W, taken in X', is $\sigma_S(X', X)$-bounded, therefore (a) is equivalent to (f). Observe that C' is a strict $\mathscr{B}$-cone in $(X', \sigma_S(X', X))$ and that each positive $\sigma(X', X)$-bounded subset of X' is $\sigma_S(X', X)$-bounded; it then follows that (f) and (g) are equivalent. It is known from lemma (7.2) that a barrel in $(X, C, \mathscr{P})$ absorbs all order-bounded subsets of X if and only if it absorbs all relative uniform null-sequences in X, then (a) is equivalent to (e). It is clear that (a) implies (b); we therefore complete the proof by showing that (b) implies (a). Let W be any barrel in $(X, C, \mathscr{P})$ which absorbs all order-bounded subsets of X. By making use of proposition (11.3), the solid kernel $\mathrm{sk}(W)$ of W is a solid barrel in $(X, C, \mathscr{P})$, hence $\mathrm{sk}(W)$ is a $\mathscr{P}$-neighbourhood of 0 by the statement (b), and therefore W is a $\mathscr{P}$-neighbourhood of 0; consequently $(X, C, \mathscr{P})$ is order-infrabarrelled.

For any locally convex Riesz space $(X, C, \mathscr{P})$, $\beta_{|\sigma|}(X, X')$ denotes the topology on X which is determined by the family

$$\{B^0 : B \text{ is } \sigma_S(X', X)\text{-bounded}\},$$

where B^0 is the polar of B, taken in X. It is clear that $\mathscr{P}$ is coarser than $\beta_{|\sigma|}(X, X')$.

(16.4) COROLLARY. *A locally convex Riesz space $(X, C, \mathscr{P})$ is order-infrabarrelled if and only if $\mathscr{P}$ coincides with $\beta_{|\sigma|}(X, X')$, and this is the case if and only if $\mathscr{P}$ is the topology of uniform convergence on the convex, solid, $\sigma_S(X', X)$-bounded sets in X'.*

Proof. This follows from theorem (16.3).

(16.5) COROLLARY. *For any order-infrabarrelled Riesz space $(X, C, \mathscr{P})$, the dual cone C' of C is a strict $\mathscr{B}$-cone in $(X', \sigma(X', X))$ if and only if it is a $\mathscr{B}$-cone in $(X', \sigma(X', X))$.*

Proof. By making use of theorem (16.3), each positive $\sigma(X', X)$-bounded subset of X' is $\mathscr{P}$-equicontinuous, the result now follows from proposition (4.8).

A locally convex Riesz space $(X, C, \mathscr{P})$ is called a *barrelled Riesz space* if the locally convex space $(X, \mathscr{P})$ is barrelled.

The following result gives many examples of order-infrabarrelled Riesz spaces.

14

(16.6) COROLLARY. *Barrelled Riesz spaces are order-infrabarrelled, and order-infrabarrelled Riesz spaces are infrabarrelled. Further, a locally convex Riesz space equipped with the order-bound topology is always order-infrabarrelled.*

Example (15.9) shows that a bornological Riesz space, and hence an infrabarrelled Riesz space is, in general, not order-infrabarrelled; while example (16.1) indicates that the class of all barrelled Riesz space is properly contained in the class of all order-infrabarrelled Riesz spaces. Now the topology of an order-infrabarrelled Riesz is the topology of a bornological Riesz space, as Nachbin (1954) and Shirota (1954) have shown.

Before giving other characterizations for which a locally convex Riesz space is order-infrabarrelled, we require the following notation: let $(X, C, \mathscr{P})$ be a locally convex Riesz space with the topological dual X' and let $X''_{|\sigma|} = (X', C', \sigma_S(X', X))'$. It is known from corollary (6.5) that $X''_{|\sigma|}$ is the ℓ-ideal in $(X')^b$ generated by X. If we define the mapping $e_{|\sigma|}$ of X into $X''_{|\sigma|}$ by putting

$$e_{|\sigma|}(x)(f) = f(x) \quad \text{for all } f \text{ in } X'$$

and if $C''_{|\sigma|}$ is the positive cone in $X''_{|\sigma|}$ consisting of all positive $\sigma_S(X', X)$-continuous linear functionals on X', then $e_{|\sigma|}$ is an injective ℓ-homomorphism of (X, C) into $(X''_{|\sigma|}, C''_{|\sigma|})$. Moreover, $e_{|\sigma|}$ is a relative open mapping of $(X, C, \mathscr{P})$ into $(X''_{|\sigma|}, C''_{|\sigma|}, \beta(X''_{|\sigma|}, X'))$; namely, for any circled convex $\mathscr{P}$-neighbourhood V of 0, $I_{|\sigma|}(V)$ is a relative neighbourhood of 0 in $I_{|\sigma|}(X)$ induced by $\beta(X''_{|\sigma|}, X')$.

(16.7) THEOREM. *For any locally convex Riesz space $(X, C, \mathscr{P})$, the ℓ-homomorphism $e_{|\sigma|}$ of $(X, C, \mathscr{P})$ into $(X''_{|\sigma|}, C''_{|\sigma|}, \beta(X''_{|\sigma|}, X'))$ is continuous if and only if $(X, C, \mathscr{P})$ is an order-infrabarrelled Riesz space.*

Proof. Let B be any $\sigma_S(X', X)$-bounded set in X'. Then the polar $B^0(X''_{|\sigma|})$ of B, taken in $X''_{|\sigma|}$, is a $\beta(X''_{|\sigma|}, X')$-neighbourhood of 0 in $X''_{|\sigma|}$. If $e_{|\sigma|}$ is continuous, then $e^{-1}_{|\sigma|}(B^0(X''_{|\sigma|}))$ is a $\mathscr{P}$-neighbourhood of 0 in X. If B^0 denotes the polar of B, taken in X, it follows from

$$e^{-1}_{|\lambda|}(B^0(X''_{|\sigma|})) = B^0$$

that B is $\mathscr{P}$-equicontinuous; and hence from theorem (16.3) $(X, C, \mathscr{P})$ is order-infrabarrelled.

Conversely, let V'' be any $\beta(X''_{|\sigma|}, X')$-neighbourhood of 0 in $X''_{|\sigma|}$, and let B be a $\sigma_S(X', X)$-bounded subset of X' such that $B^0(X''_{|\sigma|}) \subseteq V''$, where $B^0(X''_{|\sigma|})$ is the polar of B, taken in $X''_{|\sigma|}$. Since $(X, C, \mathscr{P})$ is order-infrabarrelled then, in view of theorem (16.3), B is $\mathscr{P}$-equicontinuous, and so the polar B^0 of B, taken in X, is a $\mathscr{P}$-neighbourhood of 0. The continuity of $e_{|\sigma|}$ follows from

$$e_{|\sigma|}(B^0) = B^0(X''_{|\sigma|}) \cap e_{|\sigma|}(X)$$

and
$$B^0 = e_{|\sigma|}^{-1}(B^0(X''_{|\sigma|}) \cap e_{|\sigma|}(X)) \subseteq e_{|\sigma|}^{-1}(V'').$$

Therefore the proof is complete.

Since $X''_{|\sigma|}$ is the ℓ-ideal in $(X')^b$ generated by X, it follows from theorem (13.1) that $X''_{|\sigma|}$ is the order-completion of X if and only if $X' \subseteq X^b_n$.

(16.8) COROLLARY. *For any locally convex Riesz space $(X, C, \mathscr{P})$, the following statements are equivalent:*

(a) $(X, C, \mathscr{P})$ *and* $(X''_{|\sigma|}, C''_{|\sigma|}, \beta(X''_{|\sigma|}, X'))$ *are topologically isomorphic and ℓ-isomorphic under the mapping $e_{|\sigma|}$;*

(b) $(X, C, \mathscr{P})$ *is an order-infrabarrelled order-complete Riesz space and $X' \subseteq X^b_n$.*

Proof. (b) $\Rightarrow$ (a): $e_{|\sigma|}$ is certainly an ℓ-isomorphism and an open mapping from $(X, C, \mathscr{P})$ onto the subspace $e_{|\sigma|}(X)$ of

$$(X''_{|\sigma|}, C''_{|\sigma|}, \beta(X''_{|\sigma|}, X')).$$

Further, by (b), $e_{|\sigma|}(X)$ is order-complete, and hence coincides with its order-completion $X''_{|\sigma|}$ (the condition $X' \subseteq X^b_n$ is used here (cf. theorem (13.1)). In other words, $e_{|\sigma|}$ is onto $X''_{|\sigma|}$. Finally, since X is order-infrabarrelled, it follows from theorem (16.7) that $e_{|\sigma|}$ is continuous and consequently topologically isomorphic to $e_{|\sigma|}(X) = X''_{|\sigma|}$.

In view of theorems (13.1) and (16.7), the implication (a) $\Rightarrow$ (b) is trivial.

In order to give another characterization of order-infrabarrelled Riesz spaces in terms of the closed-graph theorem, we recall the following well-known terminology: If $(E, \mathscr{T})$ and $(F, \mathscr{I})$ are topological vector spaces and if T is a linear mapping of E into F, then T is said to be *nearly continuous* if, for each $\mathscr{I}$-neighbourhood U of 0 in F, the $\mathscr{T}$-closure $\overline{T^{-1}(U)}$ of $T^{-1}(U)$ is a $\mathscr{T}$-neighbourhood of 0 in E; T is said to be *nearly open* if $\overline{T(V)}$ is an $\mathscr{I}$-neighbourhood of 0 in F whenever V

is a $\mathcal{T}$-neighbourhood of 0 in E, for each $\mathcal{I}$-neighbourhood U of 0 in F. A locally convex space $(E, \mathcal{T})$ is said to be *fully complete* (or *B-complete*, a *Ptak space*) if a subspace Q of E' is $\sigma(E', E)$-closed whenever $Q \cap A$ is $\sigma(E', E)$-closed in A for each $\mathcal{P}$-equicontinuous subset A of E'. It is well known that every Fréchet space is a Ptak space.

(16.9) THEOREM. *A locally convex Riesz space $(X, C, \mathcal{P})$ is order-infrabarrelled if and only if for any Ptak space $(F, \mathcal{T})$ the following statement holds: if T is a linear mapping of X into F such that*

(a) *T is order-bounded and*

(b) *the graph of T is closed*

then T is continuous.

Proof. Suppose that T is order-bounded and that V is a circled convex $\mathcal{T}$-neighbourhood of 0 in F. Then $\overline{T^{-1}(V)}$ is a barrel in $(X, C, \mathcal{P})$ which absorbs all order-bounded subsets of X, hence T is nearly continuous provided that $(X, C, \mathcal{P})$ is order-infrabarrelled, and so the necessity follows from a well-known result due to Ptak (cf. Robertson and Robertson (1964, p. 115)).

Conversely, let W be a solid barrel in $(X, C, \mathcal{P})$, p the gauge of W, and let $J = p^{-1}(0)$. Then p is a lower semi-continuous (Riesz) semi-norm on $(X, \mathcal{P})$, thus J is a $\mathcal{P}$-closed subspace of X (in fact, J is a $\mathcal{P}$-closed ℓ-ideal in X), consequently $(X/J, p)$ is a normed vector space. Let $(Y, \tilde{p})$ denote the completion of $(X/J, p)$, and let ϕ be the quotient mapping of $(X, C, \mathcal{P})$ into $(X/J, p)$. If B is any order-bounded subset of X, there exists $\lambda > 0$ such that $p(b) \leqslant \lambda$ for all $b \in B$, it then follows that ϕ is an order-bounded linear mapping of X into $(Y, \tilde{p})$.

By the following lemma (16.10), ϕ has a closed graph. Hence, in view of the hypothesis of the sufficiency, ϕ is continuous. In particular, $\phi^{-1}(\Sigma)$ is a neighbourhood of 0 in $(X, \mathcal{P})$, where $\Sigma = \{y \in Y : \tilde{p}(y) \leqslant 1\}$. Notice that $\phi^{-1}(\Sigma) \subseteq W$; thus W must be a neighbourhood of 0 in $(X, \mathcal{P})$. This shows that $(X, C, \mathcal{P})$ is order-infrabarrelled.

(16.10) LEMMA. *Let p be a lower semi-continuous semi-norm on a locally convex space $(E, \mathcal{T})$, let $(F, \tilde{p})$ denote the completion of the normed space $(E/p^{-1}(0), p)$, and let ϕ be the quotient mapping of $(E, \mathcal{T})$ onto $(E/p^{-1}(0), p)$. Then the graph of ϕ is closed (with respect to the product topology of $\mathcal{T}$ and the norm topology $\tilde{p}$).*

Proof. Suppose that x_τ converges to x in $(E, \mathcal{T})$, and that $\phi(x_\tau)$ converges to y in $(F, \tilde{p})$. Then, for any $\varepsilon > 0$, there exists $x' \in E$ such that $\tilde{p}(y - \phi(x')) \leqslant \varepsilon/2$. The lower semi-continuity of p with respect to $\mathcal{T}$ and the continuity of $\tilde{p}$ with respect to the norm topology on F show that

$$\tilde{p}(\phi(x) - \phi(x')) = p(x - x') \leqslant \lim \inf p(x_\tau - x')$$

$$= \lim_\tau \inf \tilde{p}(\phi(x_\tau) - \phi(x')) = \tilde{p}(y - \phi(x')).$$

We conclude from

$$\tilde{p}(y - \phi(x)) \leqslant \tilde{p}(y - \phi(x')) + \tilde{p}(\phi(x') - \phi(x)) \leqslant \varepsilon$$

that $y = \phi(x)$, and hence that the graph of ϕ is closed. This completes the proof.

It should be noted that if in theorem (16.9) we omit (b), we obtain a characterization for $\mathcal{P}$ to be the order-bound topology (cf. corollary (7.4)), while if we omit (a) and keep (b), we obtain a characterization of barrelled Riesz spaces, in view of lemma (16.10).

We conclude this chapter with a result about the topological dual of order-infrabarrelled Riesz spaces.

(16.11) THEOREM. *For any order-infrabarrelled Riesz space $(X, C, \mathcal{P})$, the topological dual X' of X is a normal subspace of X^b; consequently X' is complete for $\sigma_S(X', X)$ and also for $\beta(X', X)$. Further, $(X', C', \beta(X', X))$ is boundedly order-complete.*

Proof. Let f_τ in C' be such that $f_\tau \uparrow f$ in X^b for some $f \in X^b$. It is required to show that f belongs to X'. Observe that

$$f(u) = \sup f_\tau(u) \quad \text{for any } u \text{ in } C.$$

It follows that $\{f_\tau\}$ is $\sigma_S(X', X)$-bounded, and hence from theorem (16.3) that $\{f_\tau\}$ is an $\mathcal{P}$-equicontinuous subset of X'. In view of the Alaoglu–Bourbaki theorem, $\{f_\tau\}$ has a $\sigma(X', X)$-cluster point, say g, in X'. Since $f_\tau \uparrow$ and since C' is $\sigma(X', X)$-closed, it follows that $f_\tau \uparrow g$, and hence from $f_\tau \uparrow f$ that $g = f$. Therefore $f \in X'$ and X' must be a normal subspace of X^b.

The completeness of X' for $\sigma_S(X', X)$ and also for $\beta(X', X)$ is then a direct consequence of corollary (13.10) and proposition (13.17), and the bounded order-completeness of $(X', C', \beta(X', X))$ follows from theorem (11.16) because $(X, C, \mathcal{P})$ must be infrabarrelled.

17

PERMANENCE PROPERTIES OF
ORDER-INFRABARRELLED RIESZ SPACES

WE have seen from corollaries (15.4) and (15.7) that ℓ-ideals in a bornological Riesz space are also bornological with respect to the relative topology, and those in an infrabarrelled Riesz space are infrabarrelled with respect to the relative topology; but example (15.9) shows that this is not true for barrelled and order-infrabarrelled Riesz spaces. We shall see below that if there are some additional conditions about some sort of completeness, then the hereditary property is still satisfied for barrelled and order-infrabarrelled Riesz spaces.

We recall that an ℓ-ideal J in a Riesz space (X, C) is a σ-normal subspace of X if it follows from $0 \leqslant u_n \uparrow u$ in X with $u_n \in J$ for all natural numbers n that $u \in J$.

(17.1) THEOREM. *Let $(X, C, \mathscr{P})$ be an order-infrabarrelled, σ-order-complete Riesz space, and let J be a σ-normal subspace of X. Then J is order-infrabarrelled with respect to the relative topology.*

Proof. Let V be a solid barrel in the subspace J and suppose that
$$U = \{x \in X: \ y \in V \text{ whenever } 0 \leqslant y \leqslant |x| \text{ and } y \in J\}.$$
Then U is a $\mathscr{P}$-closed convex solid set in X such that $U \cap J = V$. Further, U must be absorbing in X. Otherwise, there exists an element x of C which fails to be absorbed by U. Hence, for each positive integer n, there exists $y_n \in J$ such that $0 \leqslant y_n \leqslant \frac{1}{n} x$ but $y_n \notin V$. By the σ-order-completeness of X, $y = \sup\{ny_n : n = 1, 2, ...\}$ exists in X. Since J is a σ-normal subspace of X, it is clear that $y \in J$. Thus $\{ny\}_{n=1}^\infty$ is contained in the order-interval $[0, y]$ in J and is not absorbed by V; this is absurd since V is a solid barrel in J. The contradition established shows that U absorbs every element in X; hence U is a solid barrel in X. Since $(X, C, \mathscr{P})$ is order-infrabarrelled, U must then be a neighbourhood of 0 in $(X, \mathscr{P})$. Since $V = U \cap J$, it follows that V is a neighbourhood of 0 in the subspace J. This shows that J is order-infrabarrelled in its own right.

(17.2) PROPOSITION.   *Let $(X, C, \mathscr{P})$ and $(Y, K, \mathscr{T})$ be locally convex Riesz spaces, and let T be a positive continuous linear mapping of X into Y. If $(X, C, \mathscr{P})$ is order-infrabarrelled and if T is nearly open then $(Y, K, \mathscr{T})$ is order-infrabarrelled.*

Proof. Let V be any barrel in $(Y, K, \mathscr{T})$ which absorbs all order-bounded subsets of Y. Then $T^{-1}(V)$ is a barrel in $(X, C, \mathscr{P})$ and absorbs all order-bounded subsets of X, so $T^{-1}(V)$ is a $\mathscr{P}$-neighbourhood of 0 in X; the near-openess of T implies that the $\mathscr{T}$-closure $\overline{T(T^{-1}(V))}$ of $T(T^{-1}(V))$ is a $\mathscr{T}$-neighbourhood of 0 in Y. We conclude from

$$\overline{T(T^{-1}(V))} \subseteq \bar{V} = V$$

that V is a $\mathscr{P}$-neighbourhood of 0 in Y. Therefore $(Y, K, \mathscr{T})$ is an order-infrabarrelled Riesz space.

As a special case of the preceding result we have the following corollary.

(17.3) COROLLARY.   *Let $(X, C, \mathscr{P})$ be an order-infrabarrelled Riesz space, and let J be a $\mathscr{P}$-closed ℓ-ideal in X. Then the quotient Riesz space $(X/J, [C]_J, \mathscr{P}_J)$ is order-infrabarrelled.*

The property of being order-infrabarrelled is preserved under the formation of inductive topologies with respect to lattice homomorphisms as the following result shows.

(17.4) PROPOSITION.   *Let (X, C) be a Riesz space and let*

$$\{(X_\alpha, C_\alpha, \mathscr{P}_\alpha): \alpha \in \Gamma\}$$

be a family of locally convex Riesz spaces. Suppose that T_α is an ℓ-homomorphism of X_α into X $(\alpha \in \Gamma)$, and that X is the linear hull of $\cup\{T_\alpha(X_\alpha): \alpha \in \Gamma\}$. If $\mathscr{P}$ denotes the inductive topology on X with respect to $\{X_\alpha\}$ and $\{T_\alpha\}$, and if each $(X_\alpha, C_\alpha, \mathscr{P}_\alpha)$ is order-infrabarrelled, then each solid barrel in $(X, C, \mathscr{P})$ is a $\mathscr{P}$-neighbourhood of 0. If, in addition, $\mathscr{P}$ is Hausdorff then $(X, C, \mathscr{P})$ is order-infrabarrelled.

Proof. Let V be any solid barrel in $(X, C, \mathscr{P})$. Since each T_α is a continuous ℓ-homomorphism of $(X_\alpha, C_\alpha, \mathscr{P}_\alpha)$ into $(X, C, \mathscr{P})$, it follows from proposition (10.24)(f) that $T_\alpha^{-1}(V_\alpha)$ is a solid barrel in $(X_\alpha, C_\alpha, \mathscr{P}_\alpha)$ and hence that $T_\alpha^{-1}(V_\alpha)$ is a $\mathscr{P}_\alpha$-neighbourhood of 0 in X_α. Consequently V is a $\mathscr{P}$-neighbourhood of 0 in X.

(17.5) COROLLARY. *The locally convex direct sum of a family of order-infrabarrelled Riesz spaces is an order-infrabarrelled Riesz space.*

Proof. Follows from corollary (11.10) and the preceding result.

It is worthwhile to remark that the quotient Riesz space and the locally convex direct sum of barrelled Riesz spaces are barrelled.

(17.6) PROPOSITION. *The completion of an order-infrabarrelled Riesz space is a barrelled Riesz space.*

Proof. Since order-infrabarrelled Riesz spaces must be infrabarrelled and since, in view of proposition (11.6), the completion of a locally convex Riesz space is also a locally convex Riesz space, it follows from a well-known result that the completion of an order-infrabarrelled Riesz space is barrelled.

(17.7) PROPOSITION. *Let $(X, C, \mathscr{P})$ and $(Y, K, \mathscr{T})$ be locally convex Riesz spaces and let T be a lattice homomorphism of X into Y. Then the following statements hold:*

(a) *if $(X, C, \mathscr{P})$ is order-infrabarrelled then T is nearly continuous;*

(b) *if $(Y, K, \mathscr{T})$ is order-infrabarrelled and if T is surjective then T is nearly open.*

Proof. (a) If V is any $\mathscr{T}$-closed, convex, solid $\mathscr{T}$-neighbourhood of 0 in Y, then $T^{-1}(V)$ is a convex solid subset of X which absorbs all order-bounded subsets of X and hence, by proposition (11.3)(a), $\overline{T^{-1}(V)}$ is a solid barrel in $(X, C, \mathscr{P})$; consequently $\overline{T^{-1}(V)}$ is a $\mathscr{P}$-neighbourhood of 0 in X. Therefore T is nearly continuous.

(b) Let U be a $\mathscr{P}$-closed convex solid $\mathscr{P}$-neighbourhood of 0 in X. Since T is surjective, it follows from proposition (10.24)(e) that $T(U)$ is a convex, solid, and absorbing subset of Y, and hence that $\overline{T(U)}$ is a solid barrel in $(Y, K, \mathscr{T})$. Therefore, in view of the hypothesis, $\overline{T(U)}$ is a $\mathscr{T}$-neighbourhood of 0 in Y and so T is nearly open.

18

RELATIONSHIP BETWEEN BARRELLED, ORDER-INFRABARRELLED, AND INFRABARRELLED RIESZ SPACES

In Chapter 15 we have given some conditions for infrabarrelled Riesz spaces to be bornological and for the topology on bornological Riesz spaces to be the order-bound topology. It is known from corollary (16.6) that locally convex Riesz spaces equipped with the order-bound topology are order-infrabarrelled and, from the example constructed by Nachbin and Shirota, that the converse is, in general, not true; therefore the following question naturally arises:

(1) Let $(X, C, \mathscr{P})$ be an order-infrabarrelled Riesz space. What condition on X (or X') is necessary and sufficient for the topology $\mathscr{P}$ to be the order-bound topology?

Since barrelled Riesz spaces are order-infrabarrelled, and since the example (16.1) shows that order-infrabarrelled Riesz spaces are, in general, not barrelled, this leads to the following question:

(2) What condition on X (or X') is necessary and sufficient for an order-infrabarrelled Riesz space $(X, C, \mathscr{P})$ to be barrelled?

Also the class of order-infrabarrelled Riesz spaces is properly contained in the class of infrabarrelled Riesz spaces, in view of corollary (16.6); therefore it is interesting for us to answer the following natural problem:

(3) What condition on X (or X') is necessary and sufficient for an infrabarrelled Riesz space $(X, C, \mathscr{P})$ to be order-infrabarrelled?

The last chapter of this book is devoted to answering these questions; we shall begin with a discussion of problems raised by question (1).

(18.1) THEOREM. *For an order-infrabarrelled Riesz space* $(X, C, \mathscr{P})$, *the following statements are equivalent:*
   (a) *$\mathscr{P}$ is the order-bound topology;*
   (b) *each order-bounded semi-norm on X is lower semi-continuous;*
   (c) *each monotone semi-norm on X is lower semi-continuous;*
   (d) *each Riesz semi-norm on X is lower semi-continuous;*
   (e) *each positive $\sigma(X^{\mathrm{b}}, X)$-bounded subset of X^{b} is $\mathscr{P}$-equicontinuous.*

Proof. The implications (a) ⇒ (b) ⇒ (c) ⇒ (d) are easy; we prove the implications (d) ⇒ (e) ⇒ (a) as follows. Suppose that the statement (d) holds. We then show that $X^b = X'$. For any $0 \leqslant f \in X^b$, let

$$p_f(x) = f(|x|) \quad \text{for any } x \text{ in } X.$$

Then p_f is a Riesz semi-norm on X, and so p_f is lower semi-continuous. Since $(X, C, \mathscr{P})$ is order-infrabarrelled, it follows from theorem (16.3) that p_f is $\mathscr{P}$-continuous. We conclude from

$$\{x \in X : p_f(x) \leqslant 1\} \subseteq \{x \in X : |f(x)| \leqslant 1\}$$

that f is $\mathscr{P}$-continuous, and hence that $X^b = X'$. On the other hand, if B is any positive $\sigma(X^b, X)$-bounded subset of X^b, then it is $\sigma_S(X^b, X)$-bounded, and so B is $\mathscr{P}$-equicontinuous in view of theorem (16.3). Therefore (d) implies (e). If the statement (e) holds, then $X^b = X'$ and $\mathscr{P}$ is the Mackey topology $\tau(X, X^b)$. Hence $\mathscr{P}$ is the order-bound topology, consequently (e) implies (a). This completes the proof.

It is known from example (15.9) that the relative topology on an ℓ-ideal induced by the order-bound topology need not be the order-bound topology. In the next result we give some sufficient conditions for this sort of hereditary property.

(18.2) PROPOSITION. *Let $(X, C, \mathscr{P})$ be an σ-order-complete, locally convex Riesz space, and let J be a σ-normal subspace of X. If $\mathscr{P}$ is the order-bound topology $\mathscr{P}_b$ then the relative topology on J induced by $\mathscr{P}_b$ is also the order-bound topology.*

Proof. Let V be a circled convex set in J which absorbs all order-bounded subsets of J. We have to show that V is a neighbourhood of 0 in the subspace J. We can assume without loss of generality that V is solid (if necessary, consider the solid kernel of V). Now, as in the proof of theorem (17.1), let

$$U = \{x \in X : \ y \in V \text{ whenever } 0 \leqslant y \leqslant |x| \text{ and } y \in J\}.$$

Then U is a convex, solid set in X such that $U \cap J = V$. Further, since X is σ-order-complete and J is a σ-normal subspace, it follows from an argument given in the proof of theorem (17.1) that U must be absorbing. Consequently, U is a circled convex set in X which absorbs all order-bounded subsets of X; hence U must be a $\mathscr{P}$-neighbourhood of 0 in $(X, \mathscr{P})$ since $\mathscr{P} = \mathscr{P}_b$. Since $V = U \cap J$, it then follows that V is a neighbourhood of 0 in the subspace J, as required to be shown.

We are now in a position to deal with the second question posed at the beginning of this chapter, namely: What condition on X (or X') is necessary and sufficient for order-infrabarrelled Riesz spaces to be barrelled? We shall see that the concept of $\mathscr{B}$-cones as well as the geometric properties of solid sets play an important role in these considerations.

(18.3) THEOREM. *Let* $(X, C, \mathscr{P})$ *be an order-infrabarrelled Riesz space with the topological dual* X'. *Then the following statements are equivalent:*

(a) $(X, C, \mathscr{P})$ *is barrelled;*

(b) *each lower semi-continuous semi-norm on* X *is order-bounded;*

(c) *each lower semi-continuous semi-norm on* X *is dominated by a lower semi-continuous Riesz semi-norm defined on* X;

(d) C' *is a* $\mathscr{B}$-cone in $(X', \sigma(X', X))$;

(e) *the solid hull of each* $\sigma(X', X)$-*bounded subset of* X' *is still* $\sigma(X', X)$-*bounded.*

Proof. The implication (a) $\Rightarrow$ (b) follows from the fact that each order-bounded subset of X is $\mathscr{P}$-bounded and that each lower semi-continuous semi-norm on a barrelled space must be continuous. By making use of theorem (16.3), (c) implies (a). Suppose now that p is a lower semi-continuous semi-norm on X, and that the statement (b) holds. For each x in X, we define

$$\tilde{p}(x) = \sup\{p(u) : 0 \leqslant u \leqslant |x|\}.$$

Since p is bounded on order-bounded sets in X, it follows that $\tilde{p}$ is finite on X. It is clear that $\tilde{p}$ is a Riesz semi-norm on X and that $p(x) \leqslant 2\tilde{p}(x)$ for all $x \in X$. If we can show that $\tilde{p}$ is lower semi-continuous, then $q = 2\tilde{p}$ is the required Riesz semi-norm. Suppose that x_τ converges to x in $(X, \mathscr{P})$ and that $\tilde{p}(x_\tau) \leqslant \mu$ for all τ and for some $\mu > 0$. For any $\varepsilon > 0$, there exists u in X with $0 \leqslant u \leqslant |x|$ such that $p(u) > \tilde{p}(x) - \varepsilon$. Suppose that $u_\tau = \inf\{u, |x_\tau|\}$, then u_τ converges to $u = \inf\{u, |x|\}$ with respect to $\mathscr{P}$, and $0 \leqslant u_\tau \leqslant |x_\tau|$. It follows from $p(u_\tau) \leqslant \tilde{p}(x_\tau) \leqslant \mu$ that $p(u) \leqslant \mu$ because p is lower semi-continuous, and hence that $\tilde{p}(x) < p(u) + \varepsilon \leqslant \mu + \varepsilon$. Therefore $\tilde{p}(x) \leqslant \mu$, and so $\tilde{p}$ is lower semi-continuous. This shows that (b) $\Rightarrow$ (c); therefore statements (a), (b), and (c) are mutually equivalent. Note that a subset B of X' is $\sigma_s(X', X)$-bounded if and only if S_B is $\sigma(X', X)$-bounded; hence by theorem (16.3), (a) $\Leftrightarrow$ (e). Also it is trivial that (e) $\Rightarrow$ (d). Conversely, suppose (d) holds. Then, by proposition (4.8), C' must be a strict $\mathscr{B}$-cone in $(X', \sigma(X', X))$. Hence if B is a $\sigma(X', X)$-bounded subset of X', then there exists an o-convex circled $\sigma(X', X)$-bounded subset A of X' such that $B \subseteq A \cap C' - A \cap C'$. Consequently the solid hull S_B of B must be contained in $2A$. This shows in particular that S_B is $\sigma(X', X)$ bounded, and hence that (d) $\Rightarrow$ (e).

Since a locally convex Riesz space equipped with the order-bound topology must be order-infrabarrelled, we record a simple consequence of the preceding theorem.

(18.4) COROLLARY. *Let the order-bound dual* X^b *of a Riesz space* (X, C) *be total over* X, *and let* $\mathscr{P}_b$ *be the order-bound topology on* X. *Then* $(X, C, \mathscr{P}_b)$ *is barrelled if and only if* C^* *is a* $\mathscr{B}$-cone *in* $(X^b, \sigma(X^b, X))$, *i.e. if and only if the conditions in theorem* (18.3) *hold.*

Before giving another characterization for order-infrabarrelled Riesz spaces to be barrelled, we need the following result.

(18.5) PROPOSITION. *Let* $(X, C, \mathscr{P})$ *be a locally convex Riesz space with the topological dual* X', *and let* $X''_{|\sigma|} = (X', C', \sigma_S(X', X))'$. *Then the following statements are equivalent:*

(a) *each* $\sigma(X', X)$-*bounded subset of* X' *is* $\sigma_S(X', X)$-*bounded;*

(b) *the topology* $\beta(X, X')$ *on* X *is the relative topology induced by* $\beta(X''_{|\sigma|}, X')$, *and the* $\sigma(X', X)$-*closure of each* $\sigma_S(X', X)$-*bounded subset of* X' *is* $\sigma_S(X', X)$-*bounded.*

Proof. (a) $\Rightarrow$ (b): Recall that the $\beta(X, X')$-topology on X is the topology of uniform convergence on the family $\mathscr{B}_1$ of all $\sigma(X', X)$-bounded subsets of X', and that the relative $\beta(X''_{|\sigma|}, X')$-topology on X is the topology on X of uniform convergence on the family $\mathscr{B}_2$ of all $\sigma_S(X', X)$-bounded subsets of X'. By (a), $\mathscr{B}_1 = \mathscr{B}_2$; it follows that the two topologies $\beta(X, X')$ and relative $\beta(X''_{|\sigma|}, X')$ must coincide on X. This proves the first assertion in (b). Further, the second assertion in (b) is a trivial consequence of (a).

(b) $\Rightarrow$ (a): Let B be any $\sigma(X', X)$-bounded set in X'. Then the polar B^0 of B, taken in X, is a $\beta(X, X')$-neighbourhood of 0 in X, and it follows from (b) that there exists a circled convex $\sigma_S(X', X)$-bounded set A in X' such that $A^0 = A^0(X''_{|\sigma|}) \cap X \subseteq B^0$, where A^0 and $A^0(X''_{|\sigma|})$ are the polars of A taken respectively in X and $X''_{|\sigma|}$. Notice that $B \subseteq B^{00} \subseteq A^{00}$. In view of the bipolar theorem, A^{00} is the $\sigma(X, X')$-closure of A; it follows from (b) that A^{00} is also $\sigma_S(X', X)$-bounded, and *a fortiori*, B is $\sigma_S(X', X)$-bounded.

As a direct consequence of proposition (18.5), theorem (16.3), and of the well-known fact that a locally convex space $(E, \mathcal{T})$ is barrelled if and only if each $\sigma(E', E)$-bounded set in E' is $\mathcal{T}$-equicontinuous, we have

(18.6) COROLLARY. *Let $(X, C, \mathcal{P})$ be an order-infrabarrelled Riesz space. Then it is barrelled if and only if the topology $\beta(X, X')$ on X is the relative topology on X induced by $\beta(X''_{|\sigma|}, X')$ and the $\sigma(X', X)$-closure of each $\sigma_S(X', X)$-bounded subset of X' is $\sigma_S(X', X)$-bounded, i.e. if and only if the conditions in theorem (18.3) hold.*

Let $(X, C, \mathcal{P})$ be a locally convex Riesz space, and let u be in C. It is easily seen that $X_u = \bigcup_n n[-u, u]$ is the ℓ-ideal in X generated by u. If p_u denotes the gauge of $[-u, u]$ on X_u and suppose that $C_u = C \cap X_u$, then (X_u, C_u, p_u) is a normed Riesz space, u is an order-unit in X_u, p_u is the order-unit norm, and hence the relative topology on X_u induced by $\mathcal{P}$ is coarser than the norm topology p_u. We shall see that the completeness of (X_u, p_u) is one of the sufficient conditions for order-infrabarrelled Riesz spaces to be barrelled (proposition (18.8)), but the completeness of (X_u, p_u) can be characterized by the fundamental σ-order-completeness of (X_u, p_u) as shown in the following.

(18.7) LEMMA. *For any locally convex Riesz space $(X, C, \mathcal{P})$ and for any u in C, the normed Riesz space (X_u, C_u, p_u) is complete if and only if it is fundamentally σ-order-complete.*

Proof. The necessity is clear. For the sufficiency, we note from theorem (8.9) that (X_u, C_u, p_u) is monotonically sequentially complete; and hence, in view of theorem (8.8) (X_u, C_u, p_u) is complete.

(18.8) PROPOSITION. *Let $(X, C, \mathcal{P})$ be an order-infrabarrelled Riesz space. For any $u \in C$, if X_u is complete for the norm p_u then $(X, C, \mathcal{P})$ is barrelled.*

Proof. Let V be any barrel in $(X, C, \mathcal{P})$, and let B be any order-bounded subset of X. There exists u in C such that $B \subseteq [-u, u]$. Since X_u is complete for p_u, then (X_u, C_u, p_u) is barrelled. It is clear that $V \cap X_u$ is a barrel in (X_u, C_u, p_u) since the relative topology on X_u induced by $\mathcal{P}$ is coarser than the norm topology p_u. Consequently

$V \cap X_u$ absorbs $[-u, u]$; in particular, V absorbs B. Therefore V is a $\mathscr{P}$-neighbourhood of 0, and thus $(X, C, \mathscr{P})$ is barrelled.

(18.9) COROLLARY. *Let* $(X, C, \mathscr{P})$ *be an order-infrabarrelled Riesz space. If* (X, C) *is* σ-*order-complete then* (X_u, C_u, p_u) *is complete for each* $u \in C$, *consequently* $(X, C, \mathscr{P})$ *is barrelled.*

Proof. Let $\{w_n\}$ be an increasing p_u-Cauchy sequence in X_u. Then $\{w_n\}$ is a p_u-bounded subset of X_u, there exists $\lambda > 0$ such that $w_n \in \lambda[-u, u]$ for all natural numbers n. By the σ-order-completeness of (X, C), $w = \sup w_n$ exists in X. Since X_u is an ℓ-ideal in X, we conclude from $-\lambda u \leqslant w \leqslant \lambda u$ that $w \in X_u$ and hence that (X_u, C_u, p_u) is fundamentally σ-order-complete. The conclusions now follow from lemma (18.7) and proposition (18.8).

(18.10) COROLLARY. *For any locally convex Riesz space* $(X, C, \mathscr{P})$ *with the topological dual* X', *if* $\mathscr{T}$ *is any locally solid topology on* X', *then* $(X', C', \mathscr{T})$ *is barrelled if and only if it is order-infrabarrelled.*

This is a direct consequence of corollary (18.9).

(18.11) COROLLARY. *A locally convex Riesz space* $(X, C, \mathscr{P})$ *is distinguished if and only if* $(X', C', \beta(X', X))$ *is order-infrabarrelled.*

Proof. It is well known (cf. Köthe 1969) that $(X, C, \mathscr{P})$ is distinguished if and only if $(X', C', \beta(X', X))$ is barrelled. The result now follows immediately from corollary (18.10).

One of the sufficient conditions for the hereditary property of barrelled Riesz spaces is easily deduced.

(18.12) COROLLARY. *Let* $(X, C, \mathscr{P})$ *be a barrelled,* σ-*order-complete Riesz space, and let* J *be a* σ-*normal subspace of* X. *Then* J *is a barrelled Riesz space with respect to the relative topology induced by* $\mathscr{P}$.

Proof. It should be noted that J is σ-order-complete. In view of theorem (17.1) and corollary (18.9), J is a barrelled Riesz space with respect to the relative topology induced by $\mathscr{P}$.

As an immediate consequence of corollaries (18.9) and (16.6), we have

(18.13) COROLLARY. *Let* X^b *be total over* (X, C), *and let* $\mathscr{P}_b$ *be the order-bound topology on* X. *If* X *is* σ-*order-complete, then* $(X, C, \mathscr{P}_b)$ *is a barrelled Riesz space.*

We shall seek some classes of locally convex Riesz spaces $(X, C, \mathscr{P})$ for which (X_u, p_u) is complete for any $u \in C$.

(18.14) PROPOSITION. *For any locally convex Riesz space* $(X, C, \mathscr{P})$, *if* $(X, C, \mathscr{P})$ *is fundamentally σ-order-complete then* (X_u, p_u) *is complete. If, in addition,* $(X, C, \mathscr{P})$ *is order-infrabarrelled then* (X_u, p_u) *is barrelled.*

Proof. It is enough to verify that (X_u, C_u, p_u) is fundamentally σ-order-complete. Let $\{w_n\}$ be an increasing p_u-Cauchy sequence in X_u. Then $\{w_n\}$ is an increasing $\mathscr{P}$-Cauchy sequence in X and $w_n \in \lambda[-u, u]$ for some $\lambda > 0$. It follows from the fundamental σ-order-completeness of $(X, C, \mathscr{P})$ that $w = \sup w_n$ exists in X. It is clear that $w \in \lambda[-u, u]$. On the other hand, since X_u is an ℓ-ideal in X, we conclude that $w \in X_u$, and hence that (X_u, C_u, p_u) is fundamentally σ-order-complete. The result now follows from lemma (18.7) and proposition (18.8).

(18.15) COROLLARY. *For any locally convex Riesz space* $(X, C, \mathscr{P})$, *if* $(X, C, \mathscr{P})$ *is monotonically sequentially complete then* (X_u, p_u) *is complete for each* $u \in C$. *If, in addition,* $(X, C, \mathscr{P})$ *is order-infrabarrelled then* (X_u, p_u) *is barrelled.*

Proof. Since C is $\mathscr{P}$-closed, the result now is a direct consequence of lemma (8.6) and proposition (18.14).

(18.16) PROPOSITION. *Let* $(X, C, \mathscr{P})$ *be a locally convex Riesz space, and let* u *be in* C. *Then* (X_u, C_u, p_u) *is complete if and only if it is monotonically sequentially complete.*

Proof. Since C_u gives an open decomposition in (X_u, p_u), the result follows from theorem (8.8).

We now turn our attention to the third question posed at the beginning of this chapter, that is: what condition on X (or X') is necessary and sufficient for infrabarrelled Riesz spaces to be order-infrabarrelled? We shall see that some sort of completeness plays an important role in these considerations.

(18.17) THEOREM. *For any infrabarrelled Riesz space* $(X, C, \mathscr{P})$, *the following statements are equivalent:*
  (a) $(X, C, \mathscr{P})$ *is order-infrabarrelled;*

(b) *each lower semi-continuous Riesz semi-norm on X is topologically bounded;*

(c) *each positive $\sigma(X', X)$-bounded subset of X' is $\beta(X', X)$-bounded;*

(d) *$(X', C', \sigma(X', X))$ is boundedly order-complete;*

(e) *$(X', C', \sigma_S(X', X))$ is both boundedly order-complete and locally order-complete;*

(f) *X' is complete for $\sigma_S(X', X)$;*

(g) *each positive $\sigma(X', X)$-bounded subset of X' which is directed upwards has a $\sigma(X', X)$-limit;*

(h) *X' is a normal subspace of X^b.*

Proof. The equivalence of (a), (b), and (c) follows from proposition (15.6) and theorem (16.3), and the equivalence of (d)–(h) follows from corollary (13.10). In view of theorem (16.11), (a) $\Rightarrow$ (h). If $(X, C, \mathscr{P})$ is infrabarrelled then, by proposition (13.15), (h) $\Rightarrow$ (a). Therefore the proof is complete.

Since each bornological space is infrabarrelled, we obtain the following corollary.

(18.18) COROLLARY. *For any bornological Riesz space $(X, C, \mathscr{P})$, if it satisfies one (and hence all) of (b)–(h) in theorem (18.17), then $(X, C, \mathscr{P})$ is order-infrabarrelled.*

The following result can be proved by a similar argument to that given in the proof of proposition (18.5).

(18.19) PROPOSITION. *Let $(X, C, \mathscr{P})$ be a locally convex Riesz space with the topological dual X', and let $X''_{|\sigma|} = (X', C, \sigma_S(X', X))'$. Then the following statements are equivalent:*

(a) *each $\sigma_S(X', X)$-bounded subset of X' is $\beta(X', X)$-bounded;*

(b) *the topology $\beta(X''_{|\sigma|}, X')$ on $X''_{|\sigma|}$ is the relative topology induced by $\beta(X'', X')$, and the $\sigma_S(X', X)$-closure of each $\beta(X', X)$-bounded subset of X' is $\beta(X', X)$-bounded.*

As an immediate consequence of theorem (18.17) and the preceding proposition, we have the following corollary.

(18.20) COROLLARY. *Let $(X, C, \mathscr{P})$ be an infrabarrelled Riesz space. Then it is order-infrabarrelled if and only if the topology $\beta(X''_{|\sigma|}, X')$ on $X''_{|\sigma|}$ is the relative topology induced by $\beta(X'', X')$, and the $\sigma_S(X', X)$-closure of each $\beta(X', X)$-bounded subset of X' is $\beta(X', X)$-bounded, that is, if and only if the conditions in theorem (18.17) hold.*

Since each normed Riesz space is infrabarrelled, we obtain the following corollary.

(18.21) COROLLARY. *Let L be a normed Riesz space with the topological dual L'. If L is order-infrabarrelled then the topology $\beta(L''_{|\sigma|}, L')$ on $L''_{|\sigma|}$ is normable.*

(18.22) COROLLARY. *Let (X, C) be an order-complete Riesz space, and let X^b_n be total over X. Then $\sigma_S(X, X^b_n)$ coincides with $\beta(X, X^b_n)$ if and only if $(X, C, \sigma_S(X, X^b_n))$ is infrabarrelled.*

Proof. The condition is clearly necessary. To prove its sufficiency, observe that $(X, C, \sigma_S(X, X^b_n))' = X^b_n$ is a normal subspace of X^b. It follows from theorem (18.17) that $(X, C, \sigma_S(X, X^b_n))$ is order-infrabarrelled and hence, from corollary (18.9), that $(X, C, \sigma_S(X, X^b_n))$ is barrelled; consequently $\sigma_S(X, X^b_n)$ and $\beta(X, X^b_n)$ coincide.

Combining theorems (18.3) and (18.17) we have the following very interesting result.

(18.23) THEOREM. *For any infrabarrelled Riesz space $(X, C, \mathscr{P})$, the following statements are equivalent:*

(a) $(X, C, \mathscr{P})$ *is barrelled;*

(b) $(X', C', \sigma(X', X))$ *is boundedly order-complete and C' is a $\mathscr{B}$-cone in $(X', \sigma(X', X))$;*

(c) X' *is complete for $\sigma_S(X', X)$ and the solid hull of each $\sigma(X', X)$-bounded subset of X' is $\sigma(X', X)$-bounded;*

(d) X' *is a normal subspace of X^b and C' is a $\mathscr{B}$-cone in $(X', \sigma(X', X))$.*

Remark. The condition that X' be a normal subspace of X^b in the preceding result can be replaced by any one of the equivalent properties listed in theorem (18.17), and the condition that C' be a $\mathscr{B}$-cone in $(X', \sigma(X', X))$ can be replaced by any one of the equivalent properties listed in theorem (18.3).

(18.24) COROLLARY. *For any bornological Riesz space $(X, C, \mathscr{P})$, if it satisfies one (and hence all) of (b), (c), and (d) in theorem (18.23), then $(X, C, \mathscr{P})$ is barrelled.*

15

NOTES ON THE BIBLIOGRAPHY

Chapters 1 & 2

The results of these two chapters, in particular, theorems (1.10), (1.12), (1.15), (1.17), and (2.11) should be considered very fundamental and important for the study of the theory of ordered topological vector spaces. The positive extension problem for linear functionals was first studied by Krein and Rutman (1948); the general characterization for linear functionals admitting positive extension, as that given in theorem (1.12), is due to Namioka (1957) and Bauer (1957, 1958). Theorem (2.8) is the work of many hands, e.g. Weston (1957b), Namioka (1957), Schaefer (1966), and Bauer (1957, 1958). The generalization of the Hahn–Banach theorem, given in theorem (1.15), is due to Bonsall and is very useful for our investigation of the duality problems for ordered vector spaces. Theorem (1.17) is essentially due to Jameson (1970). Theorem (2.11), in the present form, is taken from an article of Ng and Duhoux (1973), while parts are implicitly given in earlier papers of Ng (1970), Wong (1970a) (1973a), and Duhoux (1972a). Corollary (2.12) is given by Jameson (1970) with a different proof; but see also Grosberg and Krein (1939).

Chapters 3, 4, and 5

In the study of an ordered locally convex space $(E, C, \mathscr{P})$, two conditions have played an important role in our discussion: one condition is to say that the cone is 'large' enough to give an open decomposition property and the other is to say that the cone is 'small' enough such that $\mathscr{P}$ admits a neighbourhood-base at 0 consisting of order-convex sets. These two conditions are respectively equivalent to saying that $\mathscr{P}$ is locally decomposable and locally o-convex. Krein appears to be the first one to consider locally o-convex Banach spaces, and the duality theorem (5.15) was proved in a 1939 joint paper with Grosberg (see Krein and Grosberg (1939)). The result was generalized to general locally convex spaces by Bonsall (1957), and Schaefer (1966) studied the duality of locally o-convex spaces and $\mathscr{B}$-cones. Dually, Bonsall (1955) introduced locally decomposable normed spaces and the concept was extended by Jameson (1970) (where he used the term 'open decomposition'), Wong (1973c), and Duhoux (1972a). The dual characterization of such spaces was independently obtained by Andô (1962) and Ellis (1964). The construction of the associated locally o-convex topology $\mathscr{P}_F$ is essentially due to Namioka (1957) and that of $\mathscr{P}_D$ to Wong and Cheung (1971). The dual characterizations of $\mathscr{P}_D$ and $\mathscr{P}_F$ (in particular, the dual characterization of locally decomposable spaces) are given by Ng and Duhoux (1973b).

The equivalence of (a) and (b) in theorem (3.8) is due to Klee and the (c) equivalence is due to Jameson (1970). The short proof presented here for this theorem as well as that of theorem (3.9) is taken from Ng (1973b). The concept of nearly open decomposition is due independently to Wong and Duhoux (1972b); and, in particular, theorem (3.11) and corollary (3.13) are taken from the latter. Propositions (4.1) and (4.3) are due to Wong and Cheung (1971). Theorems (5.1) and (5.4) are due to Namioka (1957) and Schaefer (1966). For other equivalent properties for normality see Riedl (1964). Theorem (5.9) is a fundamental duality result between normal cones and $\mathscr{C}$-cones; it is due to Schaefer (1966) but part (ii) is also implicitly contained in Bonsall (1957). The proof presented here is taken from Ng and Duhoux (1973), while other short proofs were also given by Wong (1970a) and Duhoux (1972a). Theorem (5.16) is dual to theorem (5.15) of Grosberg and Krein and is due to Ellis (1964) (where he assumes that the space E is complete, and Ng (1973b) observes later that the completeness is automatic from the other assumptions by applying a generalized open mapping theorem). A somewhat less strong form of theorem (5.16) was earlier obtained by Andô (1962) (where he did not calculate the constants). The proofs of theorems (5.15) and (5.16) are taken from Ng (1970) and (1973b). Another related paper: Kist (1958).

CHAPTER 6

The concept of solid sets in a general ordered vector space was introduced by Ng (1971b) and Duhoux (1972a). Theorem (6.1) seems to be new. Theorem (6.3) is a generalization of Nachbin's result (1965) on vector lattices, in part due to Wong (1973c) and Duhoux (1972a). For the Banach space case, theorem (6.12) was proved by Davies (1968); for the present form, see Ng and Duhoux (1973). Other related papers: Wong (1969a) and Wong and Cheung (1971).

CHAPTER 7

The construction of the order-bound topology $\mathscr{P}_b$ is due to Namioka (1957) and Schaefer (1966), while the dual characterization of $\mathscr{P}_b$ is given in (1972a) by Wong. Theorem (7.3), due to Wong (1972a), can be regarded as a general form for studying the continuity of positive linear mappings. Corollaries (7.6) and (7.8) are due to Schaefer; (7.7) is established by Namioka (1957) and Klee; and (7.9) was deduced by Ng (1973a). Theorems (7.10), (7.12), and (7.14) were proved by Wong (1972a), but (7.14) was earlier obtained by Namioka in (1957) in the vector lattice case.

CHAPTER 8

The study of the relationship between order completeness and topological completeness can be broken down into two stages. The first stage is to establish some sufficient conditions ensuring that the monotonically sequential completeness implies the completeness; this had been done by Jameson (1970) for the metrizable case (cf. theorem (8.8)). The second stage is to establish some sufficient conditions ensuring that the order completeness implies the monotonically sequential completeness. This has been done by Wong for the metrizable case (cf. theorem (8.9)). With the exception of several results pointed out in the text, all results in this section are taken from an article of Wong (1972b). Other related papers: Duhoux (1972a), and Ng (1972a).

CHAPTER 9

The notion of order-unit norm is essentially due to Kadison (1950) and that of base-norm to Edwards (1964) and Ellis (1964). Much of the theory developed in this section was initiated by them. In particular, Edwards is the first to note that each compact convex set can be affine-homeomorphically embedded in a Banach dual space with the w^*-topology, and establishes the duality theorem (cf. theorem (9.10)). A dual result (cf. theorem (9.8)) was given by Ellis (1964). The notion of approximate order-unit was suggested by C^*-algebra theory and was introduced by Ng (1969a); he proved theorems (9.6), (9.9), and (9.15). The concept of the L^b-condition and results from lemma (9.24) to theorem (9.28) were cited in an article of Ng (1972b). In the case of a partially ordered Banach space with closed cone, theorem (9.7) was proved independently by Asimow (1968) and Ng (1969a), and the theorem in the present form was noted in the joint paper of Ng and Duhoux (1973); in that joint paper further generalizations of some results of this section were also discussed. The implication (a) $\Rightarrow$ (b) in theorem (9.20) is a famous theorem of Riesz (1940) and Andô proves the much more difficult implication (b) $\Rightarrow$ (a). Effros (1967) calls an ordered Banach space E with closed cone a *simplex space* if the dual E' is an AL_1-space. A intrinsic characterization (equivalent to those presented in corollary (9.22)) of simplex spaces was independently given by Davies (1967) and Ng (unpublished) at about the same time in 1966, by virtue of a powerful separation theorem of Edwards (1965). A large portion of the materials presented in this section can be found in Ng (1969a).

CHAPTER 10

Most of the material of this chapter can be regarded as mathematical folklore. The solid hull and the solid kernel (absolute core in the terminology of Roberts (1952)) were introduced by Roberts (1952). The important result of proposition (10.10) concerning the basic relation between normal subspaces and order direct sums in Riesz spaces was cited in an article of Riesz (1940). The concepts of normal integrals and integrals were introduced in an article of Nakano (1950a), and so was proposition (10.17). Systematic and extensive treatments of the theory of Riesz spaces can be found in the book of Luxemburg and Zaanen (1971).

CHAPTERS 11 and 12

The early theory of Banach lattices was studied by F. Riesz, Frendenthal, Birkhoff (1961), Kakutani (1941, 1942b), Krein (1940), and Nakano (1950a), while Roberts (1952) seems to be the first to investigate the duality theory for locally convex Riesz spaces. Theorem (11.14) concerning the completeness of topological Riesz spaces is due to Nakano (195.0), and so is proposition (11.13), but the proof that we have presented here for (11.13) is due to Schaefer (1960). A part of theorem (11.16), namely that the strong dual of a locally convex Riesz space X reflects the properties of X, was introduced by Kawai (1957); while the second assertion in theorem (11.16) on the completeness of the strong dual of infrabarrelled Riesz spaces was proved by Wong (1969b), it is a generalization of Schaefer's result (1960). Most of the results in Chapter 12 are taken from the articles of Wong (1969a) and (1969b).

Other papers or books related to the subject matter of Chapter 11: Jameson (1970), Peressini (1967), Goffman (1956, 1959), Gordon (1960), and Kuller (1958).

CHAPTER 13

The equivalence of (b) and (e) in theorem (13.1) was found by Andô, and the other equivalent properties in theorem (13.1) were proved by Luxemburg and Zaanen. Kawai (1957) and Wong (1969b) found the criterion for X to be an l-ideal in X''; their results are presented in theorem (13.5). A necessary and sufficient condition for X to be a normal subspace of X'' was proved by Wong (1969c). Corollary (13.7) was cited earlier in the article of Nakano (1950a). Theorem (13.9) on the completeness for the Dieudonné topology was established by Wong (1969c), and it generalizes results of Goffman (1956), Peressini (1967), and Schaefer (1960). Corollary (13.11) is due to Peressini (1967). Results (13.15)–(13.18) are taken from an article of Wong (1973b).

CHAPTER 14

This chapter is concerned with a study of the interrelation between reflexivity and order. Corollary (14.2) is due to Ogasawara. Corollary (14.5) was cited in an article of Schaefer (1960). Theorems (14.1) and (14.6) were found by Wong (1969c).

CHAPTER 15

Kawai (1957) proved that every bornological Riesz space is the inductive limit of a family of normed Riesz spaces (cf. proposition (15.5)); also he proved corollary (15.4). (15.1)–(15.3) and propositions (15.6)–(15.10) are taken from Wong (1970b). Other related paper: Warner (1960).

CHAPTERS 16, 17, and 18

It is known that a locally convex space equipped with the finest locally convex topology is barrelled, and that the order-bound topology is the finest locally solid topology. However example (16.1), due to Ng (1971b), shows that locally convex Riesz spaces equipped with the order-bound topology may not be barrelled. The class of order-infrabarrelled Riesz spaces, on the one hand, includes the class of locally convex Riesz spaces equipped with the order-bound topology, and on the other hand, it behaves as and plays a role similar to barrelled spaces in the theory of locally convex spaces. The class was introduced and studied by Wong in (1969d) and (1973c). In particular, he gave (1969d, 1973c) various characterizations for spaces in the class, for example that presented in Chapter 16, studied (1969d) the permanence properties of such spaces, presented here in Chapter 17, and established (1969d) some interrelationship between various classes of locally convex Riesz spaces, for example that presented here in Chapter 18. Some of his work was extended in Ng (1971b) to weakly Riesz spaces.

BIBLIOGRAPHY

AMEMIYA, I. (1960). On ordered topological linear spaces, *Proceedings of the international symposium on linear spaces (Jerusalem)* 14–23.

ALFSEN, E. M. (1971). *Compact convex sets and boundary integrals.* Springer-Verlag, Berlin.

—— (1968). Facial structure of compact convex sets, *Proc. Lond. math. Soc.* **18**, 385–404.

—— and ANDERSEN, T. B. (1970). Split faces of compact convex sets, *Proc. Lond. math. Soc.* **21**, 415–42.

—— and EFFROS, E. G. (1972). Structure in real Banach spaces, *Ann. Math.* **96**, 98–173.

ANDÔ, T. (1962). On fundamental properties of a Banach space with a cone, *Pacif. J. Math.* **12**, 1163–69.

ASIMOW, L. (1968). Well-capped convex cones, *Pacif. J. Math.* **26**, 421–31.

—— (1969). Directed Banach spaces of affine functions, *Trans. Am. math. Soc.* **143**, 117–32.

BAKER, J. W. (1968). Continuity in ordered spaces, *Maths. Z.* **104**, 231–46.

BAUER, H. (1957). Sur le prolongement des formes linéaires positives dans un espace vectoriel ordonné, *C.R. hebd. Séanc. Acad. Sci., Paris* **244**, 289–92.

—— (1958). Über die Fortsetzung positiver Linearformen, *Bayer. Akad. Wiss. Math.-Nat. Kl. S.-B. 1957* 177–90.

BIRKOFF, G. (1961). *Lattice theory* (3rd ed.). American Mathematical Society, Colloquium Publications, New York.

BONSALL, F. F. (1955). Endomorphisms of a partially ordered vector space without order unit, *J. Lond. math. Soc.* **30**, 144–53.

—— (1957). The decomposition of continuous linear functionals into non-negative components, *Proc. Durham phil. Soc.* **A13**, 6–11.

BOURBAKI, N. (1961). *Topologie générale* (2e éd.). Chapter 10, Act. Sci. Ind. No. 1084. Hermann, Paris.

—— (1965). *Intégration* (2e éd.). Chapter 1–4, Act. Sci. Ind. No. 1175. Hermann, Paris.

CHEUNG, WAI-LOK (see WONG, YAU-CHUEN).

CHOQUET, G. (1969). *Lectures on analysis.* Benjamin, New York.

DAY, M. M. (1962). *Normed linear spaces.* Springer-Verlag, Berlin.

DAVIES, E. B. (1967). On the Banach duals of certain spaces with the Riesz decomposition property, *Quart. J. math., Oxford* **18**, 109–111.

—— (1968). The structure and ideal theory of the predual of a Banach lattice, *Trans. Am. math. Soc.* **131**, 544–55.

DIEUDONNÉ, J. (1951). Sur les espaces de Köthe, *J. Analyse math.* **1**, 81–115.

DIXMIER, J. (1948). Sur un théoréme de Banach, *Duke math. J.* **15**, 1057–71.

—— (1953). Formes linéaires sur un anneau d'opérateurs, *Bull. Soc. math. Fr.* **81**, 9–39.

DUHOUX, M. (see also NG, KUNG-FU). (1972a). Topologies localement solides dans les espaces vectoriels préordonnés, *Mémoires Acad. R. Belgique, 1840.*

DUHOUX, M. (1972b). Topologies semi-dirigées et espaces vectoriels préordonnés (semi-) o-infratonnelés. *Bull. Acad. R. Belgique*, LXIII, 848–61.

—— and NG KUNG-FU (1972). The duality of convex sets in ordered vector spaces, *Rapport Séminaires, Université Catholique de Louvain, Belgique*. Rapport no. 16.

—— and NG KUNG-FU (1973a). Duality and L_p-conditions in ordered vector spaces, *Bull. Soc. Math. Belgique* (to be published).

—— and NG KUNG-FU (1973b). Homeomorphic affine embedding for a class of cones, *Bull. Acad. R. Belgique* (to be published).

EDWARDS, D. A. (1964). The homeomorphic affine embedding of a locally compact cone into a Banach dual space endowed with the vague topology, *Proc. Lond. math. Soc.* **14**, 399–414.

—— (1965). Separation de fonations réelles définies sur un simplexe de Choquet. *C.R. hébd. Séanc. Acad. Sci., Paris* **261**, 2798–800.

—— (1969). On uniform approximation of affine functions on a compact convex set, *Quart. J. Math., Oxford* **78**, 139–42.

—— (1970). An extension of Choquet boundary theory to certain partially ordered compact convex sets, *Studia math.* **36**, 177–93.

—— and VINCENT-SMITH, G. (1968). A Weierstrass–Stone theorem for Choquet simplex, *Annls Inst. Fourier (Grenoble)* **18**, 261–82.

EDWARDS, R. E. (1965). *Functional analysis*. Holt, Rinehart, and Winston, New York.

EFFROS, E. G. (see also ALFSEN, E. M.). (1967). Structure in simplexes, *Acta math.* **117**, 103–121.

ELLIS, A. J. (1964). The duality of partially ordered normed linear spaces, *J. Lond. math. Soc.* **39**, 730–44.

—— (1965). Perfect order ideals, *J. Lond. math. Soc.* **40**, 288–94.

—— (1968). On partial orderings of normed spaces, *Math. scand.* **23**, 123–32.

—— (1969). On faces of compact convex sets and their annihilators, *Math. Annls.* **184**, 19–24.

FREMLIN, D. H. (1967). Abstract Köthe spaces I, *Proc. Camb. phil. Soc. maths. phys. Sci.* **13**, 653–60.

FUCHS, L. (1966). *Riesz vector spaces and Riesz algebras* (Queen's paper in pure and applied mathematics).

GILLMAN, L. and JERISON, M. (1962). *Rings of continuous functions*. Princeton.

GOFFMAN, C. (1956). Compatible seminorms in a vector lattice, *Proc. natn. Acad. Sci. U.S.A.* **42**, 536–8.

—— (1959). Completeness in topological vector lattices, *Am. math. Mon.* **66**, 87–92.

GORDON, H. (1960). Topologies and projections on Riesz spaces, *Trans. Am. math. Soc.* **94**, 529–51.

—— (1964). Relative uniform convergence, *Math. Annln.* **153**, 418–27.

GROSBERG, J. and KREIN, M. (1939). Sur la décomposition des fonctionelles en composantes positives, *Doklady Acad. Nauk. SSSR* **25**, 723–6.

JAMESON, G. J. O. (1970). *Ordered linear spaces*, Lecture Notes in Math. 104, Springer-Verlag, Berlin.

—— (1968). Allied subsets of topological groups and linear spaces, *Proc. Lond. math. Soc.* **16**, 135–44.

KADISON, R. (1950). A representation theory for commutative topological algebras, *Memoirs Am. math. Soc.* 7.

—— (1964). Transformations of states in operator theory and dynamics, *Topology*, 3 (suppl. 2), 177–98.

KAWAI, I. (1957). Locally convex lattices, *J. math. Soc. Japan* 9, 281–314.

KAKUTANI, S. (1941a). Concrete representations of abstract (L)-spaces and the mean ergodic theorem, *Ann. Math.* 42, 523–37.

—— (1941b). Concrete representations of abstract (M)-spaces, *Ann. Math.* 42, 994–1024.

KELLEY, J. L., NAMIOKA, I., and co-authors (1963). *Linear topological spaces.* Van Nostrand, Princeton.

KIST, J. (1958). Locally o-convex spaces, *Duke math. J.* 25, 569–81.

KLEE, V. L. JR. (1955). Boundedness and continuity of linear functionals, *Duke math. J.* 22, 263–9.

KOMURA, Y. (1964). Some examples on linear topological spaces, *Math. Annln.* 153, 150–62.

KÖTHE, G. (1969). *Topological linear spaces I.* Springer-Verlag, Berlin.

KREIN, M. (see also GROSBERG, J.) (1940). Sur la décomposition minimale d'une fonctionnelle linéaire un composantes positive, *Doklady Acad. Nauk SSSR* 28, 18–22.

—— and RUTMAN, M. A. (1948). Linear operators leaving invariant a cone in a Banach space, *Usp. mat. Nauk* (N.S.) 3 No. 1 (23), 3–95 (Russian). (Also *Am. math. Soc. Transl.* No. 26 (1950).)

KULLER, R. G. (1958). Locally convex topological vector lattices and their representations, *Mich. math. J.* 5, 83–90.

LINDENSTRASS, J. (1964). Extension of compact operators, *Memoirs Am. Math. Soc.* 48.

LUXEMBURG, W. A. J. and ZAANEN, A. C. (1971). *Riesz spaces I.* North-Holland, Amsterdam.

MARTIN, J. T. (1970). Topological representation of abstract L_p-spaces, *Math Annln.* 185, 315–21.

NACHBIN, L. (1965). *Topology and order.* Van Nostrand, New York.

—— (1954). Topological vector spaces of continuous functions, *Proc. natn. Acad. U.S.A.* 40, 471–4.

NAGEL, R. J. (1971). Ideals in ordered locally convex spaces, *Math. scand.* 29, 259–71.

NAKANO, H. (1950a). *Modulared semi-ordered linear spaces.* Maruzen, Tokyo.

—— (1953). Linear topologies on semi-ordered linear spaces, *J. Fac. Sci. Hokkaido Univ.* 12, 87–104.

—— (1950b). *Modern spectral theory.* Maruzen, Tokyo.

NAMIOKA, I. (1957). (See also Kelley, J.) Partially ordered linear topological spaces, *Memoirs Am. math. Soc.* 24.

NG, KUNG-FU (see also DUHOUX, M.) (1969a). The duality of partially ordered Banach spaces, *Proc. Lond. math. Soc.* 19, 269–88.

—— (1969b). A note on partially ordered Banach spaces, *J. Lond. math. Soc.* 1, 520–24.

—— (1970). On a computation rule for polars, *Math. scand.* 26, 14–16.

NG, KUNG-FU (1971a). A duality theorem on partially ordered normed spaces, *J. Lond. math. Soc.* **3**, 403–404.

—— (1971b). Solid sets in ordered topological vector spaces, *Proc. Lond. math. Soc.* **22**, 106–120.

—— (1971c). On a theorem of Dixmier, *Math. scand.* **26**, 14–16.

—— (1972a). On order and topological completeness, *Math. Annln.* **196**, 171–6.

—— (1973a). A continuity theorem for order-bounded linear functionals, *Math. Annln.* (to be published).

—— (1972b). L_p-conditions in partially ordered Banach spaces, *J. Lond. math. Soc.* **25**, 387–94.

—— (1973b). An open mapping theorem, *Proc. Camb. phil. Soc.* (to be published).

—— and DUHOUX, M. (1973). The duality of ordered locally convex spaces, *J. Lond. math. Soc.* (to be published).

PERESSINI, L. (1967). *Ordered topological vector spaces*, Harper and Row, New York.

—— (1961). On topologies in ordered vector spaces, *Math Annln.* **144**, 199–223.

PHELPS, R. R. (1966). *Lectures on Choquet's theorem.* Van Nostrand, New York.

RIEDL, J. (1964). Partially ordered locally convex vector spaces and extensions of positive continuous linear mappings, *Math. Annln.* **157**, 95–124.

RIESZ, F. (1940). Sur quelques notions fondamentales dens la théorie générale des opérations linéaires, *Ann. Math.* **41**, 174–206.

ROBERTS, G. T. (1962). Topologies in vector lattices, *Proc. Camb. phil. Soc. maths. phy. Sci.* **48**, 533–46.

—— (1964). Order continuous measures, *Proc. Camb. phil. Soc. maths. phy. Sci.* **60**, 205–207.

ROBERTSON A. P. and ROBERTSON, W. J. (1964). *Topological vector spaces.* Cambridge University Press.

SCHAEFER, H. H. (1966). *Topological vector spaces.* Macmillan, New York.

—— (1960). On the completeness of topological vector lattices, *Mich. math. J.* 303–309.

SHIROTA, T. (1954). On locally convex vector spaces of continuous functions, *Proc. Japan Acad.* **30**, 294–8.

STØRMER, E. (1968). On partially ordered vector spaces and their duals with applications to simplexes and C^*-algebras, *Proc. Lond. math. Soc.* **18**, 245–65

VINCENT-SMITH, G. F. (see also EDWARDS, D. A.). (1973). Filtering properties of wedges of affine functions (to be published).

WARNER, S. (1960). Bornological structures, *Illinois J. Math.* **4**, 231–45.

WESTON, J. D. (1957a). Convergence of monotone sequences in vector spaces, *J. Lond. math. Soc.* **32**, 476–7.

—— (1957b). The decomposition of a continuous linear functional into non-negative components, *Math. scand.* **5**, 54–6.

—— (1957c). A topological characterization of L-spaces, *J. Lond. math. Soc.* **32**, 473–6.

—— (1959). Relations between order and topology in vector spaces, *Quart. J. Math., Oxford* **10**, 1–8.

WILS, W. (1971). The ideal center of partially ordered vector spaces, *Acta math.* **127**, 41–77.

WONG, YAU-CHUEN (1969a). Locally o-convex Riesz spaces, *Proc. Lond. math. Soc.* **19**, 289–309.

—— (1969b). Local Dedekind–completeness and bounded Dedekind-completeness in topological Riesz spaces, *J. Lond. math. Soc.* **1**, 207–212.

—— (1969c). Reflexivity of locally convex Riesz spaces, *J. Lond. math. Soc.* **1**, 725–32.

—— (1969d). Order-infrabarrelled Riesz spaces, *Math. Annln.* **183**, 17–32.

—— (1970a). A short proof of Schaefer's theorem, *Math. Annln.* **184**, 155–6.

—— (1970b). The order-bound topology on Riesz spaces, *Proc. Camb. phil. Soc.* **67**, 587–93.

—— (1972a). The order-bound topology, *Proc. Camb. phil. Soc.* **71**, 321–7.

—— (1973a). A note on open decomposition, *J. Lond. math. Soc.* (2)6, 419–20.

—— (1973b). A note on completeness of locally convex Riesz spaces, *J. Lond. math. Soc.* (2)6, 417–8.

—— (1972b). Relationship between order-completeness and topological completeness, *Math. Annln.* **199**, 73–82.

—— (1973c). Open decompositions on ordered convex spaces, *Proc. Camb. phil. Soc.* (to be published).

—— and CHEUNG, WAI-LOK (1971). Locally absolute-dominated spaces, *United College J., Hong Kong.* 241–249.

ZAANEN, A. C. (see LUXEMBURG, W. A. J.).

INDEX